W9-BIG-779

America's Favorite

Backyard Wildlife

BY
KIT AND GEORGE HARRISON

SIMON AND SCHUSTER   NEW YORK

10 9 8 7 6 5 4 3 2 1

Library of Congress Cataloging in Publication Data
Harrison, Kit.
  America's favorite backyard widlife.

  Bibliography: p.
  Includes index.
  1. Urban fauna—United States. I. Harrison,
George H. II. Title.
QL155.H37 1985   591.909′733   85-2320
ISBN: 0-671-50967-5

All of the photographs appearing in this book were taken by the authors except those appearing on the following pages:

Bill Dyer: page 296.

Hal H. Harrison: pages 17, 18, 49, 71, 95, 97, 116 *(bottom)*, 121, 124, 133, 158 *(bottom)*, 171, 184, 185, 193 (2), 194 (2), 195 (3), 198, 199, 204, 212, 226, 227, 232, 246 *(bottom)*, 261, 266, 268, 271, and 274.

(Continued on page 320)

ACKNOWLEDGMENTS

"Nothing we do, however virtuous, can be done alone" is a great Reinhold Niebuhr quote which expresses our feelings about the creation of this book.

Though it required months of research, writing and editing, and drew upon years of experience and photography, we didn't do it alone.

We are grateful to the following for their help:

John Strohm, Editor of *National Wildlife* and *International Wildlife* magazines, to whom this book is dedicated, is an old friend with whom we have been down many trails over the past twenty years. We thank him for writing the introduction.

Photographs are vital to a wildlife book. Most in this one are our own, but the others are the works of talented photographers Hal H. Harrison, Leonard Lee Rue III and his associate Irene Vandermolen, Karl and Steve Maslowski, Charles and Elizabeth Schwartz, and Bill Dyer. These accomplished nature photographers have made a major contribution to this book, and we thank them.

To wildlife artists Ned Smith and Michael James Riddet, we are indebted for the excellent line drawings contained in each chapter.

The text was read for biological accuracy by Richard M. DeGraaf, Principal Research Wildlife Biologist, the U.S. Forest Service, Amherst, Massachusetts, and by wildlife artist Michael James Riddet and his wife Karen, of Gays Mills, Wisconsin, both of whom are wildlife information specialists.

Though the authors of many fine books and research papers are cited in the bibliography, two deserve special recognition: Leonard Lee Rue III, who wrote *The World of the White-tailed Deer* and *The Deer of North America*, and Professor Lawrence Wishner, author of *Eastern Chipmunks, Secrets of Their Solitary Lives.*

A special thanks to Nancy Frank, director of the Wildlife ARC (Animal Rehabilitation Center) of Wisconsin, for helping us photograph raccoons, gray squirrels, cottontail rabbits and opossums, and to Duane Manthei, an excellent photographer himself, who did the darkroom work on a great many of our photographs.

To our staff, Ruth Utschig, Virginia Harrison and Caren Hanson, who helped us and suffered through the birth pains of this book, we are grateful.

And without the faithful support and frequent encouragement of our editor at Simon and Schuster, Dan Johnson, there would be no book.

Finally, to other friends and relatives who were most understanding during the final months of production, we say thank you.

KIT AND GEORGE HARRISON

To John Strohm, Editor,
National Wildlife and *International Wildlife* magazines,
our mentor, whose faith in us
has been important to our careers

CONTENTS

PHOTO SECTION FOLLOWS PAGE 96

Wildlife managers and research biologists, like squirrel expert Vagn Flyger, have had a significant impact on almost all wildlife populations, including those which live in backyards.

INTRODUCTION

"Why is wildlife important?" I've been asked that question many times since I put together the first issue of *National Wildlife* magazine 23 years ago.

My response is quick: "Is rain important to you? The fact that the sun comes up in the morning? Do you think of soil as just *dirt*? Do you ever see it as a source of the food we eat? Would you like to live where there are no trees? No grass?"

We take all these things for granted, but they are a vital part of our lives. Wildlife is, too. Without it, we would truly have the "silent spring" about which Rachel Carson wrote.

When I was a boy down on the farm on the horseweed bottoms of the Wabash, we had all kinds of wildlife. We had opossums, raccoons, rabbits, foxes, quail, squirrels and songbirds of all kinds. And there were wildflowers and mushrooms to hunt, the fun of finding a bee tree, and trying to beat the squirrels to the pecans and hickory nuts.

I had to work like the very dickens, driving a team of mules when I was only eight years old. Farm work was never easy, so I'm afraid I took wildlife for granted. But it was a rich heritage I'd like to bequeath to everyone.

Since that's not possible, the next best thing is to encourage you to enjoy the wildlife around your home.

Wildlife is closer than you think—as you will find from reading this book by Kit and George Harrison. The book will also entertain you with fascinating anecdotes about the wildlife in your backyard and about how the animals survive . . . eat, sleep, stay warm or cool, protect themselves and reproduce.

George came to work with me when we had only one other editor on *National Wildlife* magazine. That was 20 years ago. And Kit joined us at *Wildlife* some years later. They were both eager beavers then, and still are.

George is a crack naturalist, not just by observing, but by formal education at Penn State University. He has always been a keen observer of nature of all forms, from the hummingbird to the condor, from the ground squirrel to the mighty grizzly. He inherited from his dad a love of photography and has photographed wildlife in 43 countries around the world.

Kit also has had a great interest in wildlife. When she was only seven, she studied the activities of a 13-lined ground squirrel in the lawn of the parkway across from her home. She'd climb trees to get a better look at bird nests. At the family's lake cottage, she became familiar with redwings, martins, swallows, frogs and fishes. In her high school biology class, she helped reluctant students dissect their frogs after doing her own and finding how fascinating its anatomy was.

George and Kit make up the most knowledgeable photographic-writing team on nature that I know of. A visit with them at their home in Wisconsin is as much fun as going on a wildlife safari.

If you can't do that, the next best thing is the thrill of seeing the wildlife in your own backyard—through their eyes.

If you haven't already done so, you surely will want to make backyard wildlife a part of your life after reading this delightful, informative book.

JOHN STROHM, Editor
National/International Wildlife

FOREWORD

We are living during a period when there is an unparalleled abundance of backyard wildlife. At no time in the history of this country have so many people lived in such close proximity to ordinarily wary, secretive wild animals. In most urban, suburban and rural housing developments, wildlife has literally taken up residence at our doorsteps.

There are several reasons for this proliferation of backyard wildlife. First, a great many people have discovered how easy it is to attract wildlife to their backyards. Through *National Wildlife* and other magazines and books and television programs on the subject, millions of people have planned and planted their yards specifically for wildlife . . . and are getting results.

The National Wildlife Federation has certified thousands of backyards in the United States as "Backyard Wildlife Habitats." Owners of these yards have landscaped with the right kinds of trees, shrubs and other plants to provide both cover and natural food that attract wildlife. They have also set up feeding stations and wildlife watering areas.

As the wildlife has responded, people have invested more and more in birdseed, feeders, houses, books, binoculars and cameras and film with which to photograph their favorite backyard wildlife.

According to the latest survey conducted by the U.S. Fish and Wildlife Service, we have become a nation of wildlife watchers as some 83 million people in this country have taken up the sport. That's one out of every three!

Once on the verge of extinction, white-tailed deer are now so common in North America that they frequently become regular backyard visitors.

The second reason why there is such an abundance of backyard wildlife is that the science of wildlife management has come of age. Born in a period of desperation when a frightening number of important species were on the verge of extinction, the techniques of wildlife management required a half century to attain substantial results.

Naturalist Ernest Thompson Seton wrote in 1909, "In those, which I now call my woodchuck days, the bear, the deer, the beaver, the wolf, and even the porcupine were gone. . . ."

Instead of the hit-and-miss operations of the 1930s and 1940s, private, state, and federal wildlife managers have had a significant impact on almost all wildlife populations, including those which live in backyards. The white-tailed deer and the Canada goose are perfect examples. Fifty years ago, these two species were rare, if not candidates for an endangered species list had there been one.

Today, there are so many of both species that they are causing enormous problems in communities all over North America. Most wildlifers, however, perceive this as good news, not bad.

Finally, and perhaps most important, Americans are now turning back to nature and are finding a genuine fulfillment in their close association with the wildlife in their own backyards. They are discovering how enjoyable it is to watch the antics of gray squirrels at their bird feeders, the care a mother skunk gives her kits, the grace and beauty of a butterfly flitting across their gardens. There is something pure and honest and real about the wildlife they are seeing on the other side of the window.

And the other side of the window is where the wild things should remain. We do not advocate taming backyard wildlife, nor making them pets. In most cases, it's illegal. Keep the wild things wild, for your sake and for theirs. By all means enjoy them, but enjoy them from a distance.

This book was written for all those people who want to learn more about the secret lives of their favorite backyard wildlife . . . what they eat, where they sleep, how they reproduce, the care they give their young, how they protect themselves against the weather and their enemies, how long they live and much more.

This book was also written for those who don't have wildlife in the backyard, but would like to.

To make wildlife a part of one's daily routine is easy, inexpensive and very rewarding. By enjoying the wildlife in your backyard you will greatly enhance your own enjoyment of life . . . we guarantee it.

KIT AND GEORGE HARRISON

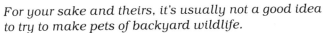

For your sake and theirs, it's usually not a good idea to try to make pets of backyard wildlife.

The ever-busy chipmunk can be a backyard delight as well as a destructible pest.

ONE

CHIPMUNK
A Cheerful Chatterbox

"The chipmunks are out!" I announced to Kit over the office intercom.

I had just seen a tawny streak shoot across the patio, followed a few inches behind by another. The two balls of fur stopped abruptly, flipped over in midair and then shot back across the patio.

It was March, and the chipmunks were back at it, searching for mates, defending homesteads and filling their pouches with sunflower seeds from the bird feeders.

From early spring until they go underground for the winter, chipmunks are ensconced outside my office window. If I happen to be out there filling bird feeders, gardening or adding water to the pond, chipmunks are always in attendance, chattering, scurrying, scrambling, romping, dashing about, occasionally brushing my trouser legs as they flit past.

The 21 species of chipmunks in North America are among the most common and adored of all species of backyard wildlife. These devilish charmers are fussy, frisky dynamos which bring a great deal of enjoyment to people all across America.

SIGN OF SPRING

Because spring is not official at our house until we see the first eastern chipmunk, it is always a major event on our backyard wildlife calendar.

Over the past decade that date has varied from February 27 to March 23. Depending on the weather, some years they are out for a few sunny days and then go back to sleep until the next sunny day. Their first appearance of the year, however, has always preceded ice-out on our lake by several weeks, yet it coincides, almost to the hour, with the arrival of the first robin.

CHARMING DECIMATORS

Chipmunks are beautifully marked, delightfully charming characters. Conversely, they are rodents, almost rats in striped clothing (biologically and figuratively). Typically, they can charm the dickens out of you and decimate your gardens at the same time.

I remember only too well the first year we had a decent strawberry crop in our backyard garden. To protect the ripening berries from the wildlife, we spread a bolt of cheesecloth over the patch. This was effective in keeping the birds from eating the precious red gems, but not the chipmunks. If only there were such a thing as a "scaremunk,"

Chipmunks can charm the dickens out of you and decimate your gardens at the same time.

One man watched a chipmunk stuff its cheeks with more than 145 kernels of grain before it occurred to him that he was being burglarized.

I would have used it. Those rascals didn't even wait for the strawberries to ripen. As soon as the berries began to show a little red, the chippies slipped under the cheesecloth and attacked without mercy. They would either pick the berries whole and carry them (it was really funny to watch them carrying berries larger than their heads), or they would take a big bite out of the center. We didn't eat more than a quart of strawberries that summer, and we finally plowed the patch under after several more frustrating years of chipmunk damage.

Another gripe: Every spring we decorate our patio borders with colorful geraniums, impatiens, and marigolds. But just as fast as we plant them, the chipmunks excavate them. And all this damage is done while they are successfully captivating us.

Chipmunks can even work their magic on hardened farmers who spend a great deal of time and money controlling mice and rat populations. One such farmer in Michigan was described in a *National Wildlife* article by Jean George as a man who had no sympathy for rats, mice, crows and squirrels when it came to protecting his granary. He would shoot any of them on sight. However, one day he became so enamored of a chipmunk in his wheat bin stuffing its cheeks with grain that he found himself counting the intake, rather than shooting the culprit. He got to 145 kernels before he realized he was being burglarized while he watched. Still, he could not shoot. Later he told George, "I just stood there and enjoyed being taken."

INDOOR RIPOFF ARTISTS

One of the most foolish things we ever allowed a chipmunk to do was enter our house. But we allowed it because we were captivated by the savvy of one particular fellow. (It happened during a major construction project a couple of years ago that required us to move out of the house.) Though this particular chipmunk's burrow was in front of the house, it figured out that the workmen left the back door open all day. It also learned that we were temporarily storing birdseed in the living room in the front of the house. The fun part was to watch Chippy run the full length of the house on the outside, enter the back door, run the full length of the house indoors (sometimes right through the legs of the workmen), gather a full cheekload of birdseed, then reverse the whole routine past the amused workmen and out the back door. That went on for a couple of weeks, until we realized that the little scamp was into more than the birdseed. It was gnawing at everything we had stored in that room. From that point on, the back door was kept shut.

We are not the first generation to deal with the fascinating behavior of this character. Chipmunks have been pulling the wool over the eyes of people for centuries. The naturalist John Burroughs permitted a chippy to take 5 quarts of hickory nuts, 2 quarts of chestnuts, and a quantity of shelled corn, just because he was curious about the animal.

The chipmunk's talent for mesmerizing people while stealing them blind can be attributed to several endearing chipmunk traits. First is their high energy level. Chipmunks are fun to watch as they scurry around and chase each other in their "Keystone Cops" routines. And they are meticulously clean animals, evidenced by their fastidious grooming behavior. Finally, they are amusing hoarders and frequently demonstrate that behavior by stuffing their cheeks to the point of looking ridiculous. It all adds up to "cute."

A CHIPPY BY ANY OTHER NAME

The eastern chipmunk is a small, ground-dwelling, solitary, diurnal rodent. Like other rodents, it has two gnawing teeth above and two below.

Chipmunks in a hurry hold their tails vertically as they scoot through flower beds, over rock walls and into woodpiles.

To build a chipmunk, you must start with 9 to 10 inches of fur-covered energy, weighing about 3 ounces. The flattened, well-haired tail is about a third of its total length. A chipmunk in a hurry will hold its tail vertical as it scoots through flower beds, over rock walls, and into woodpiles. Its head is rather short, and its ears are rounded and flattened.

Chipmunks' grizzled orange-brown to chestnut-colored bodies are marked with five dark-brown stripes down the back. A creamy buff strip separates the dark stripes on their sides.

Their striped faces distinguish them from any other mammal in most of their range. Large black eyes are bordered on top and bottom by a short creamy buff stripe outlined above and below with dark-brown stripes. The belly fur is white.

The chipmunk's four clawed toes and thumb equip it for holding food while sitting in an upright position. It has fur-covered cheek pouches into which it can put food and other material. Its front teeth are chisel-shaped.

EASTERN IS EASTERN

The range of the eastern chipmunk is virtually all of the eastern United States, except the extreme Southeast and Florida. They also range through the eastern half of southern Canada.

NAMED BY THE INDIANS

Historical records show that chipmunks were well ingrained in American Indian lore. The Chippewas, for example, called it *chetamon*, from which "chitmunk," and finally "chipmunk," evolved. A Navajo taboo is: "Never kill a chipmunk, because it will lead you to food and water." Any backyard wildlifer can vouch for the food part!

In the late 1660s, a visiting Englishman in Massachusetts called the chipmunk a "mouse squirrel." Thoreau called it "chipping squirrel." Today's nicknames include "hackee," "grinny" and, of course, "chippy," to name a few.

The Latin name for the eastern chipmunk, *Tamias striatus*, means "striped steward," one who stores and looks after provisions, referring to its habits of hoarding food. That habit, however, has made it popular in Russia, where "chipmunk nuts" fetch a higher price on the market than those gathered by people, for while to err is human, chipmunks never give houseroom to second-class nuts.

EAT AND DRINK A LOT

The chipmunk's American diet includes acorns, hickory nuts, beechnuts, cherry pits, Juneberries, raspberries, strawberries (for sure), dogwood seeds, corn and plant bulbs (don't I know!). People who feed birds in their backyards also know that chipmunks eat anything and everything birds eat, including beef suet.

Chipmunks will also raid bird nests. Two summers ago, we heard a blood-curdling scream coming from one of our birdhouses. Fifteen minutes after the screaming had ceased, a satisfied but bloody-faced chipmunk emerged. The birds it killed were starlings. We wonder how many baby cardinals, robins and catbirds our chipmunks may eat. I am convinced that one consumed a whole family of four-day-old wrens last summer.

Some observers claim that chipmunks do not require much water. Yet, we have watched chippies drink at our pool several times a day throughout their season aboveground.

Chippies drink at our pool several times a day throughout the summer.

THE CHIP OF THE MUNK

Of course, the chipmunk "chips"; it also chucks, trills, whistles, chatters and warbles. In his amazingly thorough book *Eastern Chipmunks,* which resulted from six years of study, Lawrence Wishner writes that the *chip* is a high-pitched, birdlike sound that may be expressed once or in a regular rhythm for 30 minutes or more, at a rate of 80 to 180 chips per minute. Wishner believes that the chip means nothing to other chipmunks except "I am a chipmunk and I am here."

The *chuck,* on the other hand, which is a low-pitched chip, expresses anger, annoyance, caution or outright fear.

A *chip-trill,* which is rapidly repeated chips, is heard when the animal is startled and scampers for cover, is pounced upon, or pounces upon another. The rustle of dry leaves usually accompanies the chip-trill. It may also be used as an expression of exuberance.

The *chuck-trill* is totally different, being the product of an aggressor when it attacks another.

The high-pitched *squeal* is heard during fights, while the *whistle* is given during mating chases.

Chatter, according to Wishner, sounds like a human voice played back at a faster speed than that at which it was recorded. Chatter is heard during fights or as a direct threat, and between a mother and her young.

This vocal communication system is important to chipmunks, because they are solitary, except for a short period during the breeding season and while rearing young. Though the chipmunk lives alone, it burrows in the same general area with other chipmunks and the communication between them is basic to the chipmunk social system.

AT HOME IN MOST BACKYARDS

Perhaps the best suited of all wild mammals for life in the average American backyard, chipmunks are at home in fencerows, shrubby thickets, rock walls, woodpiles, parks, gardens and wooded suburban lots. They prefer edges where they can gather nuts and seeds from woodlands as well as fruits from the more open areas.

The average home range of an adult male is about an acre. Females require only about half that space. Home ranges overlap and

Chipmunks are at home in fencerows, shrubby thickets, rock walls, woodpiles, parks, gardens and wooded suburban lots.

change in size and shape with the seasons as food availability changes. Upon leaving their mother's burrow, juveniles may have to travel some distance to establish a home range of their own, but once settled, usually remain there for life.

Chipmunks are active from dawn to dark, in most kinds of warm weather. They tend to remain in their burrows during very windy, hot or dry weather.

They can easily climb trees, especially if there is a bird feeder to be raided, though they do not jump from one branch to another as tree squirrels do. We have regularly watched our chippies as they have climbed the trunk of an ash to a feeder containing beef suet. We hope that the suet is a suitable substitute for the flesh of baby birds.

NEAT AND CLEAN AT ALL COSTS

One of the striking features of chipmunk behavior is their apparent cleanliness. They cannot be watched for long before it becomes apparent that they spend a great deal of time grooming. This is done for cleanliness, but also as a nervous behavior in the presence of others. (After all, a chipmunk never knows when it might be photographed.)

Grooming appears to be a feat in gymnastics, as chipmunks are able to reach every part of their bodies with either mouth or paws. Even their ears are thoroughly washed with their wrists and paws, which are wetted first by licking. The tail is kept tidy by whisking it through the teeth and over the tongue in one quick motion.

As a result of all this cleanliness, Wishner reports, chipmunks have very few external parasites.

WAKE-UP TIME FOR CHIPPY

Some believe that sex motivates chipmunks to wake up and dig out of their winter beds.

Soon after they emerge from their burrows in the spring, males begin to seek females as the first of two breeding seasons begins. In the northern extremes of chipmunk range, adverse spring weather may mean that the breeding season is delayed and that only one litter can be raised that year.

However, researchers Richard H. Yahner and Gerald E. Svendsen, working in southeastern Ohio, found that mating seasons there are influenced more by photoperiod (the increased hours of daylight in the spring) than weather.

Young males will try to breed during their first spring, though they may be outcourted by older, more experienced competitors. When two males meet during the breeding season, a wrestling match ensues until one of the combatants tires and acknowledges defeat.

Females generally emerge from their winter quarters about two weeks after males, and are in heat within a few days. They, too, are capable of breeding during their first spring.

NOT A CHANCE MEETING

When a male encounters a female that is not ready to breed, a nasty battle may follow, ending with the suitor sent scurrying away. In *Mammals of North America,* V. H. Cahalane writes, "From February to the middle of March, the male chipmunk begins to go abroad regularly on pleasant days. Spring has wakened his emotions and he stops cautiously at the doorways of dens where the females are still staying. It is well for him to be circumspect . . . female chipmunks are very fussy, and many a brash male has been beaten up and tossed out when he tried to rush things."

If she is in breeding condition, a female is likely to be courted by several males at the same time. If this happens, a chase will commence in which a line of males will scurry after the female through the brush, across the dead leaves, and into rock piles at incredible speed. At times when the female requires a rest, she may drop out of the chase, leaving the males to tussle among themselves until they realize that she is no longer present. Then they will scurry off to find her.

The chases will continue until the dominant male, who is usually the strongest, is the only remaining suitor.

If she is greeted by only one male, possibly the male that has taken control of her home range, the courtship begins.

Lawrence Wishner, after many years of observation, describes the courtship as follows: "Courtship is brief, with a few trills and a bit of affectionate nudging on the part of the male, who expresses his definite intentions with rapid, vertical tail movements. At all other times, chipmunks shake their tail only horizontally. Mating occupies one to two minutes, with the male thrusting rapidly while holding the female's hips with his forepaws, and with the female concentrating on remaining stationary," he reports. "No sound is produced. After mating, the two may remain together for from 20 minutes to about two hours, grooming and eating, before the female drives the male away. Females appear to mate only once during each of the two breeding seasons," Wishner concludes.

ANOTHER CROP OF CHIPPIES

During the 31-day gestation period, the female prepares a nursery in the nesting chamber of her burrow. She also increases her intake of protein, which may explain why chipmunks may suddenly become bloodthirsty in the springtime.

The three to five baby chipmunks are born naked, blind, almost shapeless and completely helpless. They weigh about 1/10 ounce.

The mother may remain inside the burrow with her newborn for several days before coming out. When she does appear on the surface, it is for a brief period. Gradually, she spends more time afield as the young approach weaning.

During the 31-day gestation period, the female increases her intake of protein.

At about three weeks, their tiny ears open and their little bodies fur out as the distinctive stripes become obvious. At four weeks, their eyes open and they are weaned.

After five to six weeks in the burrow, the young chippies are two-thirds grown and ready to venture into the outside world and try their new teeth on such delectables as fresh garden strawberries.

For the first few days, they remain close to the natal burrow and are attended by their mother. But as the days pass, the youngsters venture farther and farther afield, finding food, establishing dominance with each other and learning the rules of chipmunk society.

As the youngsters trespass into the home ranges of neighboring adult chipmunks, they are vulnerable to abuse from the landlords. After a week or two of adventures abroad and safety at home, their own mother will run them off. She will likely begin the reproductive cycle all over again, giving birth to another litter in mid to late summer.

It is during this difficult period in the life of a young chipmunk that the youngster must seek its own home range, even if it requires traveling a considerable distance.

Wishner states that one to two young females, and occasionally a male, will settle within 50 feet of the natal burrow. Some disappear completely. Some will take over unoccupied burrows; others will dig their own in a few weeks; still others may be seen wandering and hiding food on the surface as much as several months later.

Wishner also tells us that the dispersal of the litters is accompanied by a nearly continuous vocalization from the entire adult chipmunk population, which communicates to the displaced youngsters information about population density and how far they must travel to find enough space to settle. Population densities range from 2 to 30 chipmunks per acre, depending on the availability of food and habitat. These are the harsh rules of chipmunk society.

A TIME OF HOARDING

Aside from sex (and perhaps including it), the most compelling urge of the eastern chipmunk is to hoard food for the coming winter. This behavior of filling of the larder is such a compulsion with chipmunks that they seem to engage in it from the moment they emerge in the spring to the last time they disappear down the burrow in the fall.

One of the most compelling urges of the eastern chipmunk is hoarding food for the coming winter.

Hoarding is facilitated by their enormous cheek pouches, which always seem to be filled to capacity.

Observers have reported cheek pouch loads such as the following: 31 large corn kernels (2 heaping tablespoons); 145 grains of wheat; 32 beechnuts; 60 to 70 oil sunflower seeds; 16 chinquapin nuts; 13 prune stones.

Victor Cahalane reported that a chipmunk had stored as much as half a bushel of food in one burrow.

Audubon watched a chipmunk repeatedly carry four hickory nuts, two in one cheek pouch, one in the other, and one in its mouth. The chipmunk was careful to remove the sharp points on the nut hull before stuffing them into its mouth.

The apples which fall to the ground in our backyard are too large to be packed into a chipmunk's cheek pouches, and therefore each is carried singly in the mouth. This sight is guaranteed to bring a smile to even the most sober of faces. It looks like a big green ball pulling a little chipmunk.

BIRDSEED CONVEYER

Chipmunks have no problem packing their pouches with birdseed. How often I have replenished our bird feeders with seed and then watched a single chipmunk transport every single grain of it down into its burrow. I am convinced that if I had dumped a truck-load of it there, the chipmunks would not rest until every grain was underground. They never stop to eat any of it. Their passion for transporting seed to their chambers is not based on need—more likely on greed.

Wishner told me that his chipmunks have consumed 762 pounds of sunflower seed in six years.

J. A. Allen wrote in *The American Naturalist*, "This gathering of food goes on regardless of the amount needed and usually continues until the supply is exhausted or until frost or other inclement weather conditions overtake the little steward and induce him to remain at home."

Youngsters just weaned must compete with adults for the same available food to fill their own larders. Competition is usually so keen that some youngsters will not store enough food to get them through the winter. Those with the strongest instincts, however, will survive.

THIEVES IN CAMELOT

With such strong competition for food, it should be no surprise that chipmunks steal from each other. Biologist Larry Shaffer studied this aspect of chipmunk behavior and found that "theft of stored

It's not unusual to replenish bird feeders with seed and then watch a single chipmunk transport every grain of it down its burrow.

food in burrow systems when occupants were absent conceivably could take place frequently." During 40 hours of observation at a chipmunk burrow in New York State during the first two weeks in June, Shaffer saw two thieves make food-stealing raids. The thieves entered the burrow with empty cheek pouches and returned to the surface with full cheek pouches. Each thief ran toward its own burrow with the stolen food and returned shortly, repeating the robbery.

As soon as the owner of the burrow discovered the robbery, it immediately excavated a "scatterhoard" of its own (a hoard of food hidden outside the burrow) and replenished the burrow hoard. The scatterhoard was located near the owner's burrow entrance, which allowed it to watch the entrance while it was gathering food to restock its larder. Some researchers believe that theft is so common in chipmunk society that the scatterhoards are created outside the burrow primarily for restocking stolen food.

When not gathering food for the larder, chipmunks sleep in the cool, dark environment of their burrows. They spend more time there during midsummer when the temperatures on the surface are uncomfortably warm. Since the temperature underground is significantly cooler, it's a little like staying inside an air-conditioned house on a hot day.

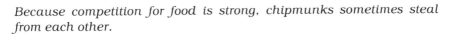

Because competition for food is strong, chipmunks sometimes steal from each other.

THE BURROW LAYOUT

One of the amazing details of the chipmunk's burrow is how perfectly clean the entrance appears. Where is all the dirt left over from the excavation?

Chipmunks are impressive earth engineers. They build their burrow network using a working tunnel entrance through which they push all the soil and rocks from the excavation. The last work is done on the main entrance, which is dug from beneath. The soil from it is used to plug up the working entrance. The location of the main entrance may change frequently for security reasons.

Naturalist Ernest Thompson Seton found a burrow entrance in the middle of his driveway in Connecticut. In an attempt to learn more about the animal, Seton plugged the entrance hole with gravel. Within 48 hours, the entrance was open again, but no gravel was evident. Seton plugged the entrance 16 more times that year and for 30 successive days the following spring. In fact, he continued his

The main entrance of a chipmunk's burrow is dug from beneath.

Soil from the tunnel and the main entrance is used to plug up the working entrance.

battle during parts of the next three years. The chipmunk always reopened the entrance, but somehow got rid of the bushels of gravel used to block it. Seton never did find the excess gravel, but conceded a victory to the chipmunk, which he said "had the satisfaction of giving a jolt to every carriage that too rudely passed his door."

The 2-inch-diameter tunnel entrance drops straight down for about 4 inches before sloping off at a 45-degree angle. In excess of 12 feet the tunnel leads to and from rooms for sleeping, storage, and latrine. Several observers have noted that chipmunks often keep different kinds of food in different chambers.

Males often build a simple network of one or two chambers, while females, requiring a nursery, will dig a more extensive facility.

Seton excavated a burrow and found a large and bulky nest in a chamber 2½ feet down. The nest measured 12 to 18 inches wide by 6 to 8 inches deep and was "warmly lined" with dry leaves.

NOT A TRUE HIBERNATOR

On that mystical day in late fall or early winter, after months of desperate hoarding and preparation, a chipmunk will at last descend the burrow, plug up the entrance, curl up into a ball and go to sleep. Though the depth of the tunnel is above the frost line, the several feet of soil between the sleeping chippy and the ice and snow on top is adequate insulation against winter's worst weather.

Chipmunks are not true hibernators like woodchucks. In Wisconsin, they disappear down their burrows sometime in late October or early November and remain there for four to five months, depending on the severity of the winter. While underground, they are torpid for only a few days to a week at a time. They awaken frequently to eat from their full larder, dispose of waste and accomplish other normal body functions deep within their winter quarters. On particularly warm winter days, a chipmunk may be seen aboveground for brief spells. With all the food they have in their pantries, they can eat when they are hungry, and thus most will emerge in excellent health the following spring.

Wirt Robinson, writing in the *Journal of Mammalogy*, records picking up a dormant chipmunk after workmen removed a boulder and exposed its burrow and nest of dried leaves. "I found inside a chipmunk, tightly coiled up, eyes closed, cold to the touch and still

In late fall or early winter, the chipmunk descends into its burrow, plugs the entrances, curls into a ball and goes to sleep.

and rigid. I moved it to another spot, placed it where the sun would strike it, and covered it with some dry leaves. Two hours later I returned and found it with its eyes open, but still stiff and unable to move. I put it in my overcoat pocket which I hung up in the warm building for an hour or so and forgot about the chipmunk. In putting on the coat later, I slipped my hand into my pocket and the chipmunk promptly bit me severely, its incisors passing through my fingernail. When I reached a suitable spot, I released it and it scampered off, now perfectly alert."

Nevertheless, not all awaken from their winter sleep. Whether it is a part of the natural scheme to control populations or an accident, we do not know. The fact remains that some chipmunks do not emerge in the spring, and many more, for various reasons, never live to see their second summer. The average chipmunk life span is slightly over a year in the wild, though they are capable of living eight to twelve years in captivity.

A PREDATOR DELICACY

The reason for the high mortality rate among chipmunks is that for many predators, chipmunk is a part of their regular diets.

Hawks, raccoons, foxes, coyotes, weasels, snakes, cats and dogs all prey upon chipmunks. Dogs and cats often make a sport of catching chippies.

Our neighbor's dog really does a job on chipmunks. Whatever Buckets' lineage is, at least some part has to be terrier. On the average, she chases chipmunks five times a day, and connects about once a week. When she catches one, she tends to play with it for a while. With the high population of chipmunks in our area, Buckets never runs out of quarry. And the exercise is undoubtedly keeping her and the chipmunks fit.

HOW TO ATTRACT CHIPMUNKS

People who want to attract chipmunks can do so by constructing rock walls with hiding spaces between the stones. They can also plant ground cover, low shrubs and food-bearing trees.

Most backyard wildlifers don't have to do anything extra to attract chipmunks. If they already have ideal habitat for birds—food, water, and a sufficient amount of cover plants, shrubs and trees—chipmunks will usually also be present.

CONTROLLING CHIPMUNKS IS HARDER

Those backyard wildlifers who feel they have too many chipmunks and want to remove them have a much greater problem. Having a chipmunk-chasing dog or cat will help, but the pet may also discourage other, more desirable kinds of wildlife. The same methods discussed in the chapter on gray squirrels will apply to chipmunks. In addition, keeping the bird food off the ground will help a great deal. Chipmunks are tree climbers, but not jumpers. They cannot jump from a tree onto a feeder as well as a gray squirrel.

Finally, don't waste your time trapping them. As I reported in *The Backyard Bird Watcher*, I trapped and transferred 46 in one

Most backyard wildlifers don't have to do anything extra to attract chipmunks.

summer and wound up with the same number—or more—in our yard. They were not the same animals, of course, but the same number of chipmunks had moved into the niches vacated by the trapped animals in a classic demonstration of a working natural vacuum.

THE OTHER CHIPMUNKS

The eastern chipmunk, *Tamias striatus*, is just one of 21 species of chipmunks in North America. They all belong to the genus *Tamias*.

The western species, which include the cliff, Colorado, Townsend, and Merriam, are all tamer, more agile and better at tree climbing than the eastern, while the eastern excavates a more extensive burrow and disappears earlier as winter approaches. The western species are more apt to hibernate in a nest located in a stump or log, or a burrow in the floor of the forest.

The least chipmunk, *Tamias minimus*, is the most widely ranging, geographically and altitudinally, of all the chipmunks. Its range overlaps that of the eastern in parts of the upper Midwest and central Canada. It is a smaller chipmunk, with stripes to the base of the tail.

Chipmunks also live in parts of Germany, northern Russia, Siberia, Mongolia, northern China, Korea, and northern Japan.

There are a couple of chipmunklike ground squirrels which also live in the western United States. The golden-mantled squirrel, *Spermophilus lateralis,* is a much larger animal, but resembles chipmunks in appearance (no stripes on face) and habits. It is frequently seen begging for food at scenic overlooks in western national parks.

The antelope squirrels of the genus *Ammospermophilus* are slimmer, with no stripes on the face, and live in the Southwest.

. . . G.H.H.

CHIPMUNK FACTS

Description: A rusty-red, fur-covered rodent with stripes on body and head, weighing about 3 ounces and 9 to 10 inches long.

Habitat: Forests, fencerows, rock walls, shrubby thickets and suburban gardens.

Habits: Diurnal, ground-dwelling squirrel. Spends most of its time in a burrow below the surface. Can climb trees. Is a passionate hoarder of seeds and nuts, which it carries in cheek pouches.

Den/Nest: Digs 12 feet or more of tunnel leading to nest, sleeping chamber and storage rooms.

Food: Omnivorous, feeding on nuts, seeds, fruit, plant bulbs, birds' eggs, young birds, insects, snails and the occasional mouse.

Voice: Chatterbox with a variety of calls including chips, chucks, trills, whistles and chatters.

Locomotion: Skitters along the ground at surprising speed.

Life Span: Average life is 1.3 years, though is biologically capable of living 12 years in the wild.

The cottontail's ears are usually standing upright, its big eyes fairly bulge, and its velvet nose is constantly awiggle.

TWO

COTTONTAIL RABBIT
Powder Puff in the Pea Patch

After a long winter, spring's sunny days have an irresistible pull for the home gardener. As soon as the danger of frost has passed, we're digging, tilling, seeding and weeding.

During one of those early-season gardening marathons a few years ago, I was raking the strawberry patch. It had been covered with straw for the winter, and now it was time to expose the plants to the sun and fresh air they needed. With enthusiastic vigor, I began raking away the straw. On the third swipe, I was startled to see that along with wilted-looking strawberry plants, I had uncovered a nest of four baby cottontail rabbits.

I called George over to share in my discovery, and we estimated that the bunnies were not quite a week old. They were well furred, and their eyes, which had been sealed at birth, were just beginning to open at the corners. We carefully replaced the warm "blanket" of fur and grass with which their mother had covered them and topped it with some straw. The strawberries would have to wait.

Cottontail nests are a fairly regular feature of our backyard, although we seldom find them. Last summer, a litter was raised in my English rock garden, just 6 feet from the back of our house. One of the youngsters continued to spend most of its time there in the clump of lavender, allowing us to watch the little rabbit develop from fist-size to full-grown.

A CLASSIC PETER COTTONTAIL

It's hard to imagine that anyone would not recognize a cottontail rabbit. From childhood, all of us have been familiar with rabbits through tales of the Easter Bunny, Peter Cottontail and Br'er Rabbit. Our earliest story books included illustrations of bunnies. Perhaps on a visit to a children's petting zoo, we squealed with delight at the chance to reach out tentatively with stubby fingers to touch the softness of their warm fur, to stroke their long ears.

The typical cottontail, we learned early in life, is covered with brown fur on its back and head and has bright-white fur on its breast and belly and the underside of its short "cottontail." Its long ears are usually standing upright, its big eyes fairly bulge, and its velvet nose is constantly awiggle. It weighs between 2 and 3 pounds and stretches out to about 18 inches.

KEEN SENSES

A rabbit's hearing is acute. With the ability to rotate its large ears, sounds can be picked up from front, back and side.

Its oversized, protruding eyes, which give the animal an innocent look, are located high on the side of the face. The rabbit has excellent vision and can see in almost all directions without having to move its head.

The busy nose seems to be ever-alert, but a rabbit's sense of smell is not as important as its vision and hearing. Wildlife photographer Leonard Lee Rue III verified this with an experience he had as a teenager on the family farm. "I was walking in the lane next to a small alfalfa field that had been harvested. This was a particularly good spot to see rabbits as there was heavy cover all around the field," he relates in *Cottontail*. "Five cottontails were eating and moving about. The air was still; the leaves hung motionless, and I followed their example. One cottontail was feeding along the path headed directly toward me. It was hop, hop, then a nibble, a taste of this, and a taste of that as though it were a gourmet's palate to be satisfied. All the while the rabbit moved closer. I hardly dared breathe. I could scarcely believe my eyes when the rabbit continued up the path, went between my legs, and continued to feed along its way. It had never noticed me, yet my scent must have saturated the air and lain heavily along the ground."

THE NATIONAL RABBIT

Cottontails of one sort or another are found in every one of the lower 48 states, up to the extreme south of Canada and down to parts of Central America. The eastern cottontail, by far the most widespread, ranges from the Atlantic Coast to Arizona.

Their habitat is every bit as varied as one might expect over such an extensive range, but the primary requisite everywhere is good cover. Cottontails find security in hedgerows, thickets, weed clumps, shrubbery, stone walls and brush piles located in swamps, meadows, backyards, pastures, marshes, cemeteries, farmlands, abandoned fields and open woodlands. They thrive in nearly every habitat from wetland to desert, provided it offers suitable cover and a ready supply of food. For rabbits, that's almost anything that's green and grows.

A SIMPLE VEGETARIAN

Grasses, sedges, leaves, flowers, herbs, stems, buds, berries and bark are the staples of a cottontail's diet.

Where green plants are not available in winter, rabbits turn to

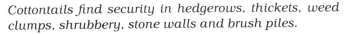

Cottontails find security in hedgerows, thickets, weed clumps, shrubbery, stone walls and brush piles.

browsing the woody stems of plants like goldenrod, barberry, snowberry, serviceberry, sumac and dogwood. A study in New York revealed that cottontails were using 71 species of plants, mostly shrubs and trees, for winter food. Orchard owners will attest that rabbits can do severe damage to young fruit trees in winter. In their desperation for food during this lean season, rabbits sometimes gnaw through the outer bark, then the inner bark of small trees, occasionally completely girdling the trunks and killing the trees.

This is the season when we are most likely to see rabbits at our bird feeders. On our front patio and in the shrubs at the back of our house we have installed feeding platforms. These are nothing more than a large square board with low sides attached to the top of a 24-to-30-inch-high tree stump. Cardinals, mourning doves, evening grosbeaks, juncos and others prefer it to the hanging feeders during the day. At night, it hosts our nocturnal visitors. Among the more common are the cottontail rabbits.

Both of these tray feeders can be illuminated with patio lights, to which the nighttime cast of raccoons, flying squirrels, opossums and cottontails soon became accustomed. Over a period of time, they also learned that our movements inside the house were not a threat to them.

If you have a similar bird-feeding setup and live where winter brings snow, you have undoubtedly seen cottontail tracks and droppings on the whiteness under the feeders on most winter mornings.

FARMER McGREGOR'S GARDEN

As soon as the first tender plant shoots poke through the ground in early spring, the cottontail begins a transition back to the succulent green things it prefers. Kentucky bluegrass is a particular favorite, along with crabgrass, alfalfa and clover.

It can't be denied that some of the tender green shoots rabbits enjoy are the ones that have been so lovingly cultivated in your own garden, with green beans being one of the all-time favorites. Cabbage is also a cottontail treat.

Stanley Rachesky of the University of Illinois Cooperative Extension Service conducted an experiment to determine which home remedies or commercially available products are effective in repelling rabbits. His test plots were planted in rows of cabbage. The plants were protected for two weeks to give them time to get established,

Grasses, sedges, leaves, flowers, herbs, stems, buds, berries and bark are the staples of a cottontail's diet.

It can't be denied that some of the tender green shoots rabbits enjoy are the ones that have been so lovingly cultivated in your garden.

then the various repellents were applied. Blood meal, mothballs, ground hot pepper, creosote-impregnated paper, jars filled with water, and two commercial products—a rabbit and a dog repellent—were used on various rows. Other rows remained untreated.

"Three domestic rabbits were placed in each section," Rachesky summarized in *Flower and Garden* magazine. "They were provided with rabbit food, water and shelter. The rabbits had the choice of feeding on the treated or untreated cabbage, or on the rabbit food provided." He said the entire test was repeated with different rabbits, and duplicate plots were also set up for wild rabbits outside the caged test plots.

"The rabbits in the cages quickly and completely ate ALL the cabbage, no matter what it was treated with," Rachesky reported. "The same results occured on the uncaged test plots."

So what's a frustrated home gardener to do? Rachesky's advice coincides with that of most authorities: "My main recommendation to prevent rabbits from feeding on your backyard plants—based on this test—is to fence your area with chicken wire buried 4 inches in the ground, with a 1x2-inch strip of wood stapled to the bottom. In our plots enclosed with such a fence, no rabbits escaped!"

Another technique came to my attention about ten years ago when I was visiting a friend in northeast Milwaukee.

"What are those jars doing in Katy's garden?" I asked Rick. My curiosity had been piqued by a dozen mason jars, each half filled with red-colored water, placed at intervals bordering the garden of Rick's elderly German neighbor.

Cottontails are regular visitors to backyard bird feeding areas, like this one on the top of a stump.

With a straight face he replied, "Katy says they keep the rabbits away."

I couldn't help smiling.

"Think what you will," he said, shrugging, "but Katy never has any rabbits in her garden. The lady across the street, however, who doesn't put jars out, loses all her bean plants to the rabbits."

Katy's mason jars stir a debate in our household to this day. To believe that they work is ridiculous, of course.

Nevertheless, I have to admit that I actually put jars around our garden for two years, fervently hoping that no one would ask why they were there. As a result—or coincidentally—we had no rabbit damage in our garden during that time. After that, I stopped using them, and, as I have already admitted, one rabbit was brazen enough to build her nest right in the middle of my garden. I can't logically admit to myself or to anyone else that those half-filled jars of colored water keep away rabbits. It's utter nonsense. And yet . . .

CRABGRASS, DANDELIONS AND RAGWEED

Another explanation for the lack of rabbit damage to our garden is more logical, if not as colorful. LeRoy Korschgen, a wildlife biologist now retired from the Missouri Department of Conservation, came up with some surprising results when he conducted a three-year food-preference study on Missouri cottontails.

Based on his examinations, rabbits prefer crabgrass over any cultivated garden plant. He found 233 plant species were consumed, but plants from only 13 families made up 95 percent of the diet. None of the 13 are grown in gardens.

The plants most often eaten by the Missouri cottontails Korschgen examined were dandelions, knotweed, daisy fleabane, lespedeza, ragweed and crabgrass.

The bark of young fruit trees rated lower than poison ivy, and garden produce was insignificant.

"Taste does play a part in the cottontails' life because, like us, they have their personal food preferences and do not necessarily always eat what is best for them," contends Lennie Rue. "It is also beyond belief what some people or creatures will eat, regardless of taste. One rabbit in Pennsylvania and another in California ate the rubber-coated ignition wires from an automobile's spark plugs. This was not an isolated instance, but a habit that these individuals developed. Each succeeded in putting a number of vehicles out of working order before being caught."

RECYCLING FOOD

Perhaps the most peculiar food habit of cottontails is coprophagy, or reingestion of some of their own droppings.

Actually, there is a similarity between this and the cud chewing among ruminants such as deer, cattle and sheep. In those animals, the cud is a mass of partly digested vegetable matter which is regurgitated, thoroughly chewed and then swallowed again. Biologists generally agree that these animals derive far more nutrients from their food when it is digested a second time before passing through the body.

In the case of the cottontail, the same is true; only the process is slightly different.

MAD AS A MARCH HARE

In early spring, cottontails turn to another important aspect of survival—reproduction.

Cottontail rabbits have been successful because of their extremely efficient and prolific breeding habits. Throughout most of the cottontail's U.S. range, the breeding season extends from March to September. Each mature female, or doe, may produce three or four litters a year, with five or six young in each.

As early as December or January, the males, or bucks, come into breeding condition. When the female comes into heat, she may be in that condition for only an hour. The males in the area sense it immediately and start to gather nearby.

The dominant male must assert his position in the social order, and, if necessary, he fights off other would-be suitors. Often the subordinate males recognize the other's dominance and simply crouch on the sidelines. If the dominant male is challenged, a real donnybrook may break out.

The two males might circle one another, waiting for an opportunity to nip at the other. If that doesn't work, they rear up on their

As early as December or January, the males come into breeding condition.

hind legs and look like small versions of boxing kangaroos. They may attempt to strike with their forepaws, but they rely on their strong hind legs and feet to deliver the meaningful blows. By leaping over his opponent, an attacking rabbit uses his hind feet to slam the challenger with a "rabbit punch." It's no wonder, with all the kicking and biting that's part of this ritual, that there's often a lot of rabbit fur floating around when it's all over.

COURTING IN LEAPS AND BOUNDS

Eventually one male will triumph and approach the female. During courtship, the two cottontails sometimes engage in what looks like a playful romp. Time and again, one or the other will suddenly leap straight up. The other dashes underneath before the jumper touches down again. What may be happening is that the female is launching attacks on the male, trying to bite his flanks. The male leaps high to get out of her way as she charges.

The actual breeding is accomplished in a matter of seconds. Then, likely as not, the female drives away the male, nipping at him and sometimes actually pulling out a mouthful of fur.

If a female is not bred, or if she does not become pregnant, she will continue to come into heat at seven-day intervals until she is successfully bred.

Rabbit courtship involves a lot of leaping, bounding and dashing.

A COZY NEST

The gestation period for cottontails is fairly short, usually less than 30 days. In preparation for the birth of her litter, the doe prepares a nest. If the nest is disturbed, or if something doesn't seem quite right, the doe will desert it and dig another. She may excavate several before she is satisfied.

The nest site is generally in an open area, not in heavy cover. It might be in a meadow, a pasture, a flower bed or even your lawn. Often the female will expand a depression that is already present, like a footprint, a small crater from a dislodged rock or the scrape of a foraging skunk. Pawing with her front feet, the doe scoops out a bowl about 5 inches deep and 5 to 7 inches in diameter. A lining of grass and dry leaves is added. Then she plucks fine, soft fur from her breast and belly and creates a cozy inner lining in the cup.

Using more grass and leaves, she creates a mat that is pulled over the top of the nest. This "blanket" will not only keep her family warm, it will also hide the nest. Construction completed, the doe leaves the nest until the birth of her young.

A QUICK DELIVERY

Labor, such as it is, seems to come upon cottontail mothers quite suddenly. Often there is no time to get to the nest before the babies arrive. The birth of an entire litter may take about 45 minutes.

If the mother is at the nest, the babies are born into it. If she delivers them away from the nest, each new cottontail is licked clean, nursed and then carried to the nest in the manner of a mother cat transporting her kitten. Bathed and fed, the litter is "tucked in." The female pulls the blanket cover over her family and leaves. She does not spend time at her nest unless she is nursing the young, but she remains in the vicinity, close enough to keep guard.

THE REPRODUCTIVE TREADMILL

After the birth of her litter, the female is again in heat, and every male in the area knows it.

"Immediately the female is pursued by the males, because her estrous period takes place at once," Rue states. "Fights between the males are more common now, and one male may charge into another trying to mount the female in an effort to dislodge him. Copulation again takes place and may involve several males. In about half an hour the males abandon the one female and seek out another."

With a breeding schedule like that, it's no wonder that cottontails are prolific.

Cottontails come into the world with eyes sealed and ears pressed tightly against their heads.

The female does not spend time at her nest unless she is nursing her young, but she remains in the vicinity, close enough to keep guard.

The nest is lined with dry grasses and soft fur plucked from the mother's breast and belly.

MEANWHILE, BACK AT THE NEST . . .

Snug in the soft fur-lined nest, the 4-inch-long newborns are completely helpless. Weighing barely an ounce each, they came into the world with eyes sealed and dainty ½-inch-long ears pressed tightly against their heads. There may be from three to eight youngsters in a litter, but four to six is typical.

During the day, their mother remains nearby in her form, a shallow depression in suitable cover where she rests and watches the nest. Should she feel that her young are at risk from a dog, fox or other predator, she may display admirable valor. Generally timorous, a female cottontail can become a terror when her offspring are threatened. She might rush out to meet the villain head-on, leaping at it and delivering blow after blow with powerful kicks from her hind legs. Other times, she might try to draw the predator's attention to her, then lead it off on a zigzag chase.

Being a crepuscular and nocturnal creature, the cottontail is most active from dusk to dawn. With the shield of twilight, the cottontail mother cautiously emerges from her hiding place and approaches the nest. With her forepaws and muzzle she pulls away the nursery covering and stretches out on top of the nest. It appears she is doing nothing more than spending a few moments at leisure. In fact, she is nursing her young. When the infants feel her presence, each pulls itself up to a nipple to feed. Afterward, the doe replaces the concealing blanket and again leaves the nest. She'll return at dawn, and perhaps once or twice during the night, to repeat the procedure. Meanwhile, she satisfies her own appetite and remains close enough to the nest to observe any interlopers.

Nature writer Alan Devoe discovered a nest of newborn rabbits on a hillside one summer. The next day, he returned with his wife to check on the bunnies. He poked aside the coverlet with his forefinger. "Mary and I looked at each other in surprise," he recalled in *Our Animal Neighbors*. "Our almost unidentifiable little pink beings of the day before had turned into rabbits overnight.

"They were still blunt-headed, lumpy and pinkish, and their little rabbit ears still lay flattened and shriveled, rather like the damp crumpled wings of a moth when it has just hatched from its cocoon," he related. "Their eyes, of course, were still sealed shut. They were still blind, deaf, helpless little mites of rabbits. But they were rabbits. They had begun to grow fur. They were only a day and a half or at most two days old, but already their vigorous little bodies—wriggling away from my finger strenuously as I touched the blanket—were shadowed with a perceptible fuzz. It behooves little cottontails to grow up fast. We had not realized, until this moment, just how fast."

By the age of one week, give or take a day, the youngsters' eyes are open and their ears are upright. Their bodies are covered with soft fur, and their foreheads are marked with a white blaze, a feature which disappears with maturity.

Sometime around two weeks after birth, young cottontails make their first excursions from the nest. They play, explore and begin weaning themselves on succulent green grasses and other tender vegetation. For a few days, they return to the nest to sleep. Then they disperse.

A cottontail's childhood is fleeting. By the age of 16 to 20 days, the youngsters are independent and their mother is probably making preparations for the birth of her next litter.

It was at this stage that we first noticed the youngster that had selected our English rock garden for its daytime hideout. One day while I was watering the aubrieta, alyssum, lavender and campanula,

At twilight, the female returns to the nest to nurse her young.

At the age of about one week, the youngsters' eyes open and their ears are upright.

A cottontail's childhood is fleeting. These young are about to become independent as their mother prepares for another litter.

a small ball of brown fur shot out from the lavender. The bunny was no larger than a man's fist, still with an endearing baby face and the overall roundness typical of very young mammals. It disappeared under the grapevine and I didn't give it another thought.

Two days later, the same thing happened, and again not long afterward. The juvenile had definitely adopted the rockery as its day-

Most cottontails are on their own by the time they are 16 to 20 days old.

time home. Realizing that this was the case, we tried not to disturb it more than necessary. Because the rock garden is directly outside our sunroom windows, it was easy to keep an eye on the young cottontail's development. It became accustomed to our voices from the other side of the windows and eventually paid no attention to us. We watched the little rabbit reach full size in a matter of weeks. The growth from just one week to the next was astounding.

John Kriz, wildlife biologist with the Pennsylvania Game Commission, does not consider a Pennsylvania cottontail to be fully grown if it weighs less than 33 ounces, or about 2 pounds.

"It takes about 16 to 20 weeks for a cottontail, on this particular area, to reach adult size," he reported. "It also appears, although data are limited, that rabbits born later in the summer reach the 2-pound size in less time than it takes the earlier born bunnies. Why? We don't know," he admitted. "Possibly it is nature's way of getting all these animals ready for winter by allowing the late litters to catch up to the early ones."

On the average, female rabbits are a few ounces heavier than the males. Kriz says the heaviest cottontail caught in his southwestern Pennsylvania study was a 60-ounce female.

There are records of cottontails breeding at the tender age of two and a half to three months, but most wait until the following spring.

FORM IS EVERYTHING

One of the first orders of business for a cottontail fresh out of the nest is to establish a form. A form is where the rabbit hides and rests during daylight hours. It is usually a shallow depression that the bunny has trampled down or scratched out in heavy cover. Typically it will be well concealed in tall grass or weeds, a brush pile, a thicket or similar dense growth, as in our rock garden.

Cottontails don't roam far from their forms, although they may use more than one form. Their home range is small, and may extend no farther than one backyard. Researchers at the University of Wisconsin found that sizes of home ranges tended to vary by season, sex and individual. Home ranges of adult females, for example, were a little over 4 acres in spring, then decreased to slightly more than 2 acres from early summer through mid-January, according to the researchers, Tracey Trent and Orrin Rongstad. By comparison, they found that the home ranges of adult males increased from about 7 acres in spring to nearly 10 acres in early summer, then dropped to less than 4 acres in late summer.

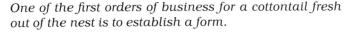

One of the first orders of business for a cottontail fresh out of the nest is to establish a form.

At times, however, a male may travel over 20 or 30 acres, ranging over the territories of several females and other males.

Whatever the size of the home range, each cottontail rabbit becomes intimately acquainted with every rock, stump, tree root, weed patch, fence post and burrow in its own territory. When chased, the rabbit is very reluctant to leave its home range for unfamiliar ground. An experienced rabbit hunter knows this, and waits in one spot while his beagle chases the rabbit in a big circle past the hunter.

Ernest Thompson Seton was intrigued by this rabbit idiosyncrasy. A female cottontail would come to his garden from the nearby open woods every evening after sundown. "One day, I saw her enter by an open gate. I closed the gate and shouted," he said. The cottontail ran straight for an opening at a point about 3 feet high in Seton's wire fence. "She leaped for that, but failed to make it; so without a moment's hesitation or doubt, she steamed across to another point 200 feet away, and leaped up and through a hole in the wire at a place about 2 feet from the ground. A woven-wire fence has not many landmarks to hold the memory," he remarked. Yet the rabbit knew her territory so well that she was aware of every escape route available to her, and made straight for them with unerring accuracy.

This is one of the cottontail's defenses against the many meat eaters that make rabbit a staple of their diet. Foxes, minks, weasels, free-roaming dogs, coyotes, skunks, bobcats, owls and the larger hawks and snakes are their major wild predators. On occasion, a shrew will kill cottontail nestlings.

PROTECTION IN STILLNESS

When it senses danger, a rabbit may simply freeze in position, relying on its brown coat for camouflage.

When flushed, a rabbit's usual tactic is to dash in a zigzag pattern, white cottontail flashing conspicuously, as it circles its familiar home territory. Then the rabbit may abruptly duck into a hiding place like a thicket or a woodchuck burrow. Or it may merely stop running and hunker down, tail pressed low, relying again on its coloration to hide it. The pursuer, zeroing in on the white tail, suddenly loses its target and may give up the chase.

In desperation, cottontails have been known to plunge into water, swimming rapidly to escape the pursuer. The rabbit that made headlines with President Jimmy Carter a few years ago was probably

When they sense danger, rabbits may simply freeze in position.

in that panic-stricken state. Perhaps forced to take to the water in a last-ditch escape effort, the rabbit frantically swam to the boat from which Carter was fishing and tried to scramble in. The President, not realizing that the rabbit was not a secret weapon launched against him, promptly dispatched the hapless intruder with an oar.

When hopelessly cornered, cottontails can strike out with their hind feet. Only in extremely rare circumstances have they been driven to bite.

HIGH MORTALITY IS THEIR LOT

These defenses work sometimes, but overall, they are not very effective. Cottontail mortality is extremely high. In an average year, more than 80 percent won't survive to their first birthday.

Many are killed while still in the nest. In spite of the best efforts of the mother, heavy rains can drown the nestlings, agricultural plows and mowers can slaughter them, and predators can dig them out and devour an entire nestful at one sitting. Studies have found that only one-third to one-half of the cottontails born ever leave the nest.

*In desperation, cottontails have been known to plunge
into water, swimming rapidly to escape the pursuer.*

DISEASES AND PARASITES, TOO

Cottontails are also susceptible to a long list of parasites and
diseases. One of the more common cottontail diseases, and one
which can be transmitted to humans, is tularemia, or rabbit fever.

Discovered 75 years ago in Tulare County, California, by Dr.
George McCoy of the U.S. Public Health Service, tularemia is a bac-
terial disease to which cottontails are highly susceptible. Rabbits
that contract tularemia are usually dead within a week.

The danger to humans comes from handling an infected animal.
Cottontail hunters are among those most likely to come in contact
with an infected animal. The disease can be picked up from the cottontail's blood or other body fluids when the hunter is in the process
of cleaning the rabbit. Usually tularemia in humans resembles a mild
case of flu. Still, it is fatal in about 5 percent of the cases. Among
rabbits, it is always fatal.

With all the hazards facing cottontails, it is hardly surprising
that their life expectancy is less than a year in the wild.

RABBIT MULTIPLICATION TABLES

These facts are not too disturbing considering the fecundity of
the cottontail rabbit. Someone once computed that if every offspring

produced by a single pair of cottontails over a five-year period survived, their progeny would number 350,000.

Even so, there are fewer cottontails today than there were a century ago. "Clean-farming practices had not become common, so that there was much more cover in which the rabbits could hide," maintains Lennie Rue. "By the last part of the 19th century, too, many marginal farms were abandoned and were allowed to grow back into brushland. All these things combined to push the number of cottontail rabbits to an all-time high, from which there has been a long steady decline that cannot be reversed," he contends.

Hunting seems to have little effect on rabbit populations. Researchers have repeatedly confirmed this. Beginning with the same number of rabbits in a control area on which hunting is not allowed and in a hunted area of equal size, they find that the end numbers consistently turn out to be close to the same. Disease and predation take a heavy toll when hunting does not, working to keep nature in balance.

As conservationist John Madson once pointed out when speaking of rabbit populations, "Warped as it sounds, it doesn't usually harm rabbits to shoot them. It may even be good for them. For a good rabbit crop, like a row of radishes or a pondful of panfish, is healthier if it's thinned out."

With all the hazards facing cottontails, it is hardly surprising that their life expectancy is less than a year in the wild.

No longer able to rely on the greenery of spring and summer, most cottontails change from grazers to browsers in winter.

Cottontails are not well equipped to travel in heavy snow.

MENU CHANGES WITH THE SEASONS

Changing weather in the autumn and winter brings a change in the cottontail's diet. No longer able to rely on the greenery of spring and summer, most cottontails must make a transition to a diet which includes wood fibers, mainly buds, bark and twigs. It turns from grazer to browser.

With snow on the ground and temperatures frigid, a cottontail may hole up for a while. It might stay in a form located in very dense cover, or it might go underground in a woodchuck, badger or skunk burrow. These dens offer some protection from certain predators and provide shelter from the elements.

Cottontails are not well equipped to travel in heavy snow like the varying, or snowshoe, hares. They are easily bogged down, and their alternative is to use tunnels in the snow. Two researchers at Purdue University believe that these snow tunnels also provide protective cover as the cottontails travel to and from feeding and resting areas.

"Tunnels through drifted snow seemingly were dug by cottontails; they occurred consistently in open areas between cover and browse concentrations," observed the researchers, Mark Fitzsimmons and Harmon P. Weeks, Jr. Reporting in the *Journal of Mammalogy*, Fitzsimmons and Weeks said that well-formed travel lanes ran to and from the snow tunnels, often with many runways converging into a single tunnel. There were signs of excavation at the ends of tunnels, they said, but they never saw the rabbits engaged in the construction of the tunnels. "One alternative explanation of tunnels is that they were formed when rabbits dug out after being covered by drifted snow. Although all tunnels had entrances at both ends, rabbits that returned to use a burrow thus formed conceivably could have dug or burst out in a different direction," they speculated.

BASICS WILL KEEP THEM HOME

Whether you live in the Far North, the Deep South, the West or anywhere in between, you'll find it's easy to attract cottontail rabbits to your backyard habitat.

The foremost consideration for these timid creatures is sufficient cover in which they will feel secure. Making large brush piles is one of the best things you can do to provide instant rabbit habitat. Thickets, dense shrubbery, grapevines and fencerows also offer the kind of hiding places rabbits need.

Food is of less importance. If you have a lawn, you've got rabbit food. They'll help keep your Kentucky bluegrass mowed, munch on your crabgrass and clean up the spilled seed from under your bird feeders.

In most areas of the country, cottontails derive sufficient water from the plant material they eat and have little need for a separate water source. In winter, they lick snow and ice.

Cottontails will clean up the spilled seed from under your bird feeders.

RABBITS, RABBITS, RABBITS

The eastern cottontail, *Sylvilagus floridanus,* and its many subspecies is the most widespread cottontail in North America. Other cottontails, like the desert cottontail, *Sylvilagus auduboni,* and the mountain cottontail, *Sylvilagus nuttali,* of the West, and *Sylvilagus transitionalis,* the New England cottontail of the Northeast, closely resemble the more common eastern cottontail.

Other close relatives include the swamp rabbit, *Sylvilagus acquaticus,* and the marsh rabbit, *Sylvilagus palustris,* both natives of Dixie.

The brush rabbit, *Sylvilagus bachmani,* and the pygmy rabbit, *Sylvilagus idahoensis,* are small rabbits of the West.

Jackrabbits and snowshoe rabbits (varying hares) are not rabbits; they are hares. Unlike the rabbit, the hare is precocial—young are born fully furred, with eyes open and ears alert, and they leave the nest soon after birth. They are ready to run the same day they are born. Hares' ears are longer than those of rabbits, and their hind legs and feet are larger and more powerful. In some, like the varying hare,

Lepus americanus, the coat turns white in winter, providing an ideal camouflage for the northern hare's winter habitat.

. . . K.P.H.

COTTONTAIL FACTS

Description: Typical "bunny." Furred in brown; belly and underside of tail are white. Big eyes; long, upright ears; stubby "cottontail."

Habitat: Meadows, pastures, backyards, parks, roadsides, hedgerows, fencerows, open woodland. Particularly likes areas providing brushy cover, thickets, tall weeds.

Habits: Generally crepuscular, most active at dawn and dusk. Usually solitary, except when breeding or in nest as young.

Den/Nest: The den, or form, is usually a shallow depression in dense cover where the animal rests during the day. In severe weather or to escape a pursuer, a cottontail may seek shelter in a woodchuck or badger burrow. The nest built by the female for her young is a shallow bowl lined and covered with grasses, leaves and soft fur which she plucks from her own body.

Food: Nearly any type of plant life. In summer, grasses, leaves, flowers and sedges are staples. In winter, bark, twigs and buds.

Voice: Generally silent. Mother may make soft grunting sounds to her young. Youngsters in nest may squeal. When facing death, can emit a loud, piercing scream. Like the howl of a timber wolf or the call of a loon, this haunting sound, once heard, is never forgotten.

Locomotion: When at ease, a cottontail may move about slowly, making movements that look like slow-motion hops. If pursued, it bounds along with impressive speed. Each leap may cover 5 to 15 feet.

Life Span: The life expectancy of a cottontail is certainly less than a year in the wild; some say as little as six months. In a few rare instances, researchers have captured the same tagged individual cottontail over a period of three or four years. In captivity, they have lived more than five years.

The jet-black robber's mask is one of the most distinguishing characteristics of this nighttime prowler.

THREE

RACCOON
Masked Rapscallion

A shadowy figure stealthily made its way over the railing and onto one of the balconies outside my second-floor office. I was still working at my desk about 9:30 P.M., well after dark, when I caught a glimpse of the intruder out of the corner of my eye.

I turned the beam from my swing-arm lamp toward the French doors. "Well! What are *you* doing there?" I asked the big raccoon that was sitting on its haunches looking into the blinding light. Its lack of response to my question was immaterial. I knew perfectly well what it was doing there. It was looking for birdseed.

From the time we get our first hard frosts in October until mild temperatures return in May, the two ledges that we have installed on my office balconies become tray seed feeders for the birds. In the warmer months, the ledges hold flower boxes full of colorful geraniums, petunias and vinca.

No doubt this coon had put my balcony feeders on its regular nightly rounds when the seed was plentiful. This was early June . . . no seeds, only flowers. It sat there in the darkness, nose pressed against the screen, looking at me. I knelt on the other side, nose pressed against the window, looking at the raccoon. It didn't seem at all bothered by my movements or my voice, so I called to George to join me and my visitor.

The raccoon, meanwhile, sniffed the air, slowly turned away and lumbered off, probably to pay a call at the platform bird feeder outside the sunroom below.

To see the coon on the balcony was no surprise; we'd been expecting it. For some time, we had noticed muddy footprints trailing up the side of the house near our back door.

After two or three more nocturnal visits, the raccoon realized that the bountiful seed buffet was no longer available at that particular site, so it eliminated my balcony from its rounds.

A ROGUISH MASKED BANDIT

This particular masked prowler was a fine specimen of a full-grown raccoon. It was approximately 30 inches long from the tip of its velvety black nose to the end of its bushy ringed tail. We estimated its weight under the woolly, grizzled brown-gray coat to be about 15 pounds. It was a good size for a raccoon, but not even close to the all-time record. That raccoon, shot by Albert Larson of Nelson, Wisconsin, in 1950, was 55 inches long and weighed 62 pounds.

The five to seven black rings on the 10-inch tail and the jet-black robber's mask over the eyes are the most distinguishing characteristics of a raccoon. That mask and the bright black eyes, perky erect ears and insatiable curiosity (not to mention appetite) combine to

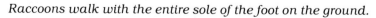

Raccoons walk with the entire sole of the foot on the ground.

make a creature that most of us find utterly endearing, if somewhat pesky.

A lumbering gait is its typical pace when it is foraging over a small area. It can burst into a run to cover ground faster between foraging points, and can certainly move with great dexterity in climbing trees or pouncing on a frog, but its maximum speed is barely 15 mph.

Raccoons, like people and bears, are plantigrades. They walk with the entire sole of the foot on the ground rather than just the toes as do digitigrades like cats, dogs and horses. Their tracks reflect this. A raccoon footprint bears a strong resemblance to the footprint of a human baby.

CHIRPS, CHURRS, SNARLS AND SQUEALS

When they are very young, raccoons learn to squeal if they are handled roughly, scream when frightened or annoyed and chirp an *err-err-err* or *orr-orr-orr* when begging for food. They whimper and cry when discontent; they purr loudly when content. The distress cry of a baby raccoon is similar to the cry of a human baby.

Juveniles and adults commonly *churr* when feeding. Another raccoon vocalization is a tremulous call which is very much like a harsh, high-pitched version of the screech owl's *whoo-oo-oo-oo*.

We heard most of these sounds clearly two years ago when we spent the summer in a friend's gatehouse while our own house was being expanded. A family of raccoons occupied a den in an old sugar maple just outside our bedroom door.

During the day we heard nothing from them, but when darkness came, the nocturnal raccoons were very active. Squalling raccoon youngsters or the squeals and screams of an interfamily brawl jolted me upright in bed on more than one occasion. The sound is like a combination of snarling dogs, fighting cats and screaming banshees.

The first—and only—night they managed to get into the garbage can, we heard their churring under our bedroom window as they dined on chicken bones, salad scraps and bits left in the bottoms of assorted cans. They had so deftly opened the garbage can and removed its contents that we didn't realize what they were feeding on until the next morning when we saw the refuse spread over a 30-foot swath. After cleaning up the mess, we did our best to "raccoon-proof" the garbage can.

PICKPOCKET DEXTERITY

Not only are raccoons adept at getting into almost any garbage can, regardless of clamps or rocks on the lid meant to keep them out, they are also skilled at figuring out how to open doors, lift latches, pull corks from bottles and open jars.

Their fingers are extremely sensitive, and everything that interests them is handled, manipulated, rubbed with the fingers. They often locate their food entirely by touch.

There is no opposable thumb like ours among the five fingers on a raccoon's forepaw, but it still is able to grasp things. "One raccoon that I had raised by bottle-feeding used to delight in going through my pockets while I held it," recalled Leonard Lee Rue III in *The World of the Raccoon*. "It would have made an excellent pickpocket because it had no trouble extracting a coin as thin as a dime. It didn't even have to bend its fingers to pick up the coins but would bring out several at a time, held separately between its fingers."

INQUISITIVE RASCALS

Most of us can't help enjoying the antics of raccoons, even if their actions are less than desirable. "No one, even the gardener with his sweet corn patch in a shambles, can remain angry with a raccoon for long," insists Missouri Department of Conservation writer Joel Vance. "They're too cute," he says.

"How can you curse a critter that, when you're scolding it, will put its paws over its eyes?" Vance asks.

There's no doubt that raccoons have personality plus. Writing in *Smithsonian*, A. B. C. Whipple told of an "indulgent host who claims to have been regularly visited by a raccoon that sits on his sofa watching television, yawning during commercials and on occasion going over to the set to switch channels." Raccoons do not care if a TV is color or black-and-white, he pointed out. They are color-blind.

He also tells of a music lover who swears that a neighborhood raccoon is a Beethoven fan. "It prefers the Ninth Symphony," Whipple claims. "When the Ninth starts booming through the open windows, the raccoon promptly comes out of the woods, unlatches the screen door, walks to one of the speakers and sits beside it, immobile and

Raccoons are adept at getting into almost any garbage can, regardless of devices meant to keep them out.

unblinking, until the last bars have sounded. Then it rises quietly, lets itself out and returns to the forest."

With traits like these, it's no wonder that many people think they'd like to have a raccoon as a pet, even though in most places, it's now illegal to keep wild animals as pets without proper licensing.

When they are young, raccoons are indeed amusing to have around, if their quarters have been "raccoon-proofed." Otherwise, they'll get into everything they can, which will include a lot of things that it would never occur to their owners to protect. However, when they reach sexual maturity, they generally become ornery and dangerous. It's time to release them back to the wild. For most animals, this would be disastrous after having become accustomed to life as a pampered pet. A raccoon will do fine. It is perhaps the only North American animal that handily makes the transition from household pet to wild animal. Turned loose, the raccoon goes back to making a living by foraging in woodland, wetland or suburb.

TREES AND STEEPLES

Because they can thrive in a diversity of habitats, raccoons have been able to establish themselves in every one of the lower 48 states, as well as north into Canada, and south far into Mexico.

Good raccoon habitat is anywhere they can find food, water and den sites.

Good raccoon habitat can be along a woodland lake or stream, in a marsh or swamp, on farmland, in suburbia or even in the inner city where they can find food, water and den sites.

When they can't find a hollow in a tree for a den in a downtown area, they improvise. Cliff Hoffman and Jack Gottschang studied a population of raccoons a few years ago in Glendale, Ohio, a suburb near Cincinnati. Their published report in the *Journal of Mammalogy* emphasized the adaptability of the raccoon in their account of a "typical" juvenile female coon that they radio-tracked. She customarily spent most of the night feeding in dumpsters behind commercial buildings. Before dawn, she'd move to a house, where she'd feed in the garbage can. Then the researchers would lose her radio signal for a while. Closer investigation showed that she was following an underground drain culvert to reach one of her daytime resting places. "Toward the end of her tracking period, a resting site was found inside the belfry of a nearby chapel," they reported.

The Glendale raccoons, like others that live in cities and suburbs, also used sewers, houses, barns, garages and sheds as dens in which to sleep during the day or bear their young in the spring. Young raccoons were annually born and raised on the roof of Glendale's Christ Church, according to Hoffman and Gottschang. Obviously, civilization has not interfered much with the raccoon's way of life. To the contrary, raccoons are thriving in suburban backyards across America.

Civilization has not interfered much with the raccoon's way of life. They are thriving in suburban backyards across America.

WINTER COURTSHIP

Their breeding season is in the dead of winter. Most mating occurs in January, February and March, with southern raccoons usually starting in January, northerners in February. Females are often ready to breed at the age of one year, but males generally aren't capable of breeding until they are two.

When the time is right, the male raccoon leaves his snug den to wander in search of a mate, checking out every spot that might harbor a receptive mate. When the polygamous male finds a female, she decides whether or not to accept him. The female seems to be monogamous; she'll mate with only one male, and she chooses which it will be. For example, if two males fight over one female, the victor may not necessarily win the favor of the lady in question.

When the male is accepted, he is likely to move into the female's den with her for a week or two of coon courtship before she drives him out and he plods off in search of another mate. He'll continue this pattern until the end of the breeding season in March.

When the courting male is accepted by a female, he is likely to move in with her for a week or two.

Having been bred, the female typically curls up for a long nap until spring. Raccoons don't hibernate, but they're usually torpid in cold weather, and in the North they sleep most of the winter.

CUBS ARRIVE IN SPRING

There is no need for the female to prepare a special nest for her cubs. They will be born in the den that has been her home den over the winter. Whatever rotted wood chips or shavings happen to be there already provide the only nesting material. She does not bring in anything else.

Her cubs, born in April or May, 63 days after conception, usually number from three to seven. Most often there are four.

As each cub emerges from the birth canal, the female licks it thoroughly, removing the embryonic sac, which she eats along with the placenta. Her tireless tongue instinctively removes the chorionic fluid from around the kit's nose and mouth, stimulates it to breathe as a smart slap on the fanny does for a newborn human baby, and helps dry the infant's short, fine fur.

A newborn raccoon is a helpless 4-inch-long mite weighing 2 to 3 ounces. Its dark skin is covered with soft, thin fur and its ears and eyes are sealed. Faintly discernible even now are the characteristic black mask and tail rings. After ten days, they are distinct.

FROM BABY TO TEENAGER

Between 18 and 23 days, the little ones' eyes open, but the youngsters remain huddled together in the den, sleeping most of the time. When their mother is away too long on her evening feeding sortie, or when they are hungry, they whimper and whine.

At the age of one month, they have grown to about 2 pounds. They are still small enough for the mother to pick up by the scruff of the neck—like a mother cat with a kitten—if her den has been disturbed and she wants to move them to new quarters.

When the mother is away, the youngsters remain huddled together in the den.

In a matter of weeks, the babies can walk, run a little and climb.

By seven weeks, they can walk, run a little and climb. "Like kittens, baby raccoons will often climb to the very tiptop branches and sit there and cry for help," says Lennie Rue.

When they are around ten weeks old, the cubs begin following their mother on her nightly forays. Cautious at first, staying close at her heels, they become bolder as the days pass. They start to eat solid foods as they begin the weaning process. Insects, berries and small aquatic animals are among their first foods.

For a few days, the family returns to the home den to sleep. Then they may wander farther on their nocturnal sojourns through their home range and hole up in whatever den is convenient when daylight approaches. A typical home range for a raccoon would be from 12 acres to an area 2 miles in diameter.

LEARNING GOOD AND BAD HABITS

Sometime between the age of 12 and 16 weeks, the youngsters are weaned and have learned how to catch crayfish and frogs, find wild berries, raid sweet-corn fields, destroy nests of birds by eating their eggs (and sometimes the young birds), break into garbage cans and help themselves to birdseed at feeders.

Our neighbors were frustrated for weeks by a young raccoon that visited their feeding station every night. It made straight for the tubular Droll Yankee filled with expensive niger seeds. The coon lifted the hanging feeder, took off the cover, raised the tube to its lips and drank down the seeds in a single draft.

Over the years, it's been a challenge for us to provide beef suet for the woodpeckers without having it available to the coons at night. We finally solved that problem with a rectangular Hyde Bird Feeder Company suet feeder made from plastic-coated hardware cloth. The hinged lid is fastened with a crimped cotter pin. We've used that feeder, wired to a nail on a tree trunk, for about six years now, and those talented raccoon fingers have yet to get into the suet. Of course, there have been occasions when they've made off with the entire feeder. The next morning, we'd find it on the hillside, perhaps scratched or gnawed a bit, but otherwise inviolate.

When they are around ten weeks old, the cubs begin following their mother on her nightly forays.

A. B. C. Whipple became exasperated with the raccoon that was finding dinner in the Whipple garbage can in Old Greenwich, Connecticut. Whipple's garbage can was recessed in a well with a step-on lid. Hearing the clang of the garbage-can lid one night, Whipple made an inspection. The lid was closed, but he could hear noises inside the can. "As I approached it, the top opened, lifted by the head of a large raccoon which nimbly climbed out, glanced my way with what I took to be a look of gratitude and sauntered away," he wrote in his *Smithsonian* account.

"In hindsight, I realize that it was a look of benign contempt," he continued. "Evidently our garbage was not up to the standards of the habitat, because on subsequent nights our visitor took to hauling the plastic bags from the can and distributing the contents about the driveway, evidently to facilitate the selection process." Whipple decided that the coon was laying out the courses of its dinner. "One morning I found a spread of shrimp shells followed by clean-picked chicken bones followed by cheese and crackers followed by a cigar butt," he reported.

He tried various techniques designed to keep the marauding raccoon out of the garbage can. "A hinge on one side of the can, a hasp and clip on the other side, a spring lock across the can top, and a hood over the rack have combined to protect our garbage—but only temporarily, I'm sure," he lamented. "Moreover, no one else in my family can open the garbage can, and the garbage collector is annoyed at the extra time it takes him to work the combination on his rounds!"

Garbage aside, a raccoon's diet is generally more than 75 percent vegetable matter and about 25 percent animal matter. Raccoons are omnivorous, eating whatever they can easily obtain. Like people, they

Sometime between the age of 12 and 16 weeks, the youngsters are weaned and have learned how to catch crayfish and frogs and find other food.

Some raccoons like to haul out the entire contents of a garbage can before making their selections.

are especially fond of certain foods. "My Flossie developed such a passion for ant eggs, used for turtle food, that we had to keep them under lock," said Bil Gilbert of his pet raccoon. "Flossie solved the problem by prying loose the aquarium cover and eating the turtles," he wrote in *Reader's Digest*. Flossie was also fond of sweets. "Like others of her kind, she had a sweet tooth, ready to sell her soul for candy, ice cream, cake, pie or almost anything else sugary," he claimed.

In the wild, raccoons are sometimes scorned by wildlife managers who know that plundering raccoons can destroy nearly half of the waterfowl nests in some prairie pothole areas.

In agricultural regions, raccoons can nearly drive farmers to tear their hair when they raid fields of almost-ripe sweet corn. Adroitly peeling back the husks, the coons chomp on the juicy kernels, sometimes just a couple of bites, sometimes eating nearly one-third down the cob. Then they go on to the next.

THE "WASH-BEAR"

When they can, raccoons forage in streams and ponds for crayfish, frogs, tadpoles, minnows and salamanders. They wade into the water, reach down and start exploring with their sensitive fingers.

Probing under rocks and in the mud, they feel rather than see what they're doing. Meanwhile, they keep their eyes on the shoreline to look for predators or anything else that happens to interest them.

Apparently water increases the fingers' sensitivity, because captive raccoons (not wild ones) often take their food to their water and dunk it. Then they manipulate it thoroughly underwater, as though in this way they will better know what they're about to eat.

For these reasons, the raccoon was named *Procyon lotor* by the early biologists. *Lotor* means "one who washes." In Germany, where the raccoon was introduced many years ago, it is called the *Waschbar*, which means "wash-bear."

PUTTING ON POUNDS FOR WINTER

Raccoons, young and old, spend the summer months eating, eating, eating. They seem to be most obvious to their human neighbors during this time of year. This is also when people are most likely to try to entice their local raccoons to make daily visits to patios and kitchen doors. Once the raccoons realize there are table scraps to be had in the same place every night, they'll be regulars. This is a very bad "hobby" in which to indulge. Problems arise if the food is no longer made available—for example, when the homeowners are out of town. Raccoons have been known to lift a door latch or pry open a window, walk into a kitchen, open the refrigerator and help themselves.

Raccoons are often scorned by wildlife managers because they destroy so many waterfowl nests.

Raccoons often forage in streams and ponds for aquatic delicacies.

By autumn, whether stuffing themselves with your table scraps or their natural wild foods, some raccoons reach weights in excess of 20 pounds. The young of the year might be 10 to 15 pounds. The entire body, even the tail, carries a thick layer of fat.

Lennie Rue once skinned and dissected a 21¼-pound raccoon. "I removed almost nine pounds of pure fat from this animal. It would be safe to conclude that in the late fall, fat makes up almost 50 percent of the animal's total weight," he contended.

SNUGGLING IN FOR WINTER

The fat helps the raccoon survive through the winter when food is scarcer and little time is spent foraging.

In the North, raccoons start denning up for the winter in November or December. In the South, they remain quite active throughout the winter months.

Some raccoon young, especially those in the North, stay in the same den with their mother until she forces them out in spring to make room for her new litter.

Others, particularly in the South, disperse when autumn arrives, striking out on their own to find a suitable den in which to spend the winter.

Natural tree cavities are the most desirable, but when these cannot be procured, a raccoon may settle underground in a woodchuck burrow. The woodchuck, deep in hibernation and sealed into one of the side chambers, neither knows nor cares that it is sharing its home with another.

In the South, young raccoons usually disperse when autumn arrives, striking out on their own to find a suitable den in which to spend the winter.

Raccoons do not hibernate. If the temperature is above about 28 degrees F., they come out of the den at night as usual. When the temperature drops below that, they remain curled up inside, nose and toes tucked under the body, tail wrapped around like a muffler. As the winter wears on, the coons apparently become acclimatized to the frigid temperatures. By January, they are often seen out and about on nights when the thermometer is at 0 degrees F.

By the time spring arrives, most coons weigh about half what they did at the end of autumn.

Over the years, coonskin caps and coats have gone in and out of fashion.

EVEN COONS FACE HAZARDS

Some young coons, conceived late in the breeding season and born a month or two later than others, have a hard time building up sufficient fat stores before winter is upon them. If the winter is severe, these youngsters may not survive until spring.

Winter death among the late-born young is just one form of raccoon mortality. Although most predators won't face the ferocity of a full-sized raccoon, bobcats, fishers, mountain lions, coyotes and wolves are capable of overpowering one. Occasionally, a great horned owl will take a youngster.

A raccoon will not seek a fight; it prefers to make a hasty exit, if possible. But if cornered, it fights fiercely. Its sharp claws can rip an attacker's undersides open or slash the jugular.

Stories abound of raccoons swimming out into a river or lake to escape dogs. When the dog follows, the coon circles behind the dog, climbs onto the dog's head and holds its pursuer underwater until the dog drowns. Or so they say. Because this story is told so often, and by so many who seem to be reliable, it may be true.

Stories abound of raccoons swimming into a river or lake to escape dogs.

Whether you like it or not, they'll also want to get in on the offerings at your bird feeders.

MAN TOPS LIST OF PREDATORS

Man is the raccoon's primary predator. He hunts the raccoon with dogs until it is treed, or traps it for its pelt. Over the years, coonskin caps and coats have gone in and out of fashion. Raccoon meat, dark and fatty, is an acquired taste, but some people think it's delicious.

Untold numbers also die on the nation's highways. But even with all the roadkills and the ups and downs of hunting and trapping pressure, best estimates put today's raccoon population about even with what it was in the mid-1800s.

YOU'VE PROBABLY GOT 'EM

Most of us who are backyard wildlifers aren't interested in the raccoon's pelt or its flesh. For us, it is a delightful, intelligent, amusing, sometimes troublesome rascal.

Many people make deliberate attempts to lure raccoons to their backyards by offering table scraps. That's the wrong way to do it; you might be very sorry in the long run.

If you have a backyard that has mature hardwood trees (especially one or two with large natural cavities), a water source nearby and available food, you probably have raccoons. Natural food for them would be the fruit of wild grapes, persimmons, fruit trees or berries, for example. Whether you like it or not, they'll also want to get in on the offerings at your bird feeders.

If there are no den trees in your backyard, try nailing a big nesting box 15 to 25 feet high in a large tree like a maple. One with somewhat larger overall dimensions than a wood-duck house would be good. If the raccoons don't use it, squirrels, flickers or others undoubtedly will.

NO LOOK-ALIKES

The raccoon is the only one of its family that is a common backyard visitor.

The coati, *Nasua nasua*, is a rare, diurnal, tropical creature of Mexico, extreme southwestern New Mexico and the southeastern quarter of Arizona. It usually travels in bands of up to a dozen and lives in open forests. Its head and body measure about 20 to 25 inches, and the long, banded tail adds another 20 to 25 inches to the animal's overall length. Its coat is grizzled brown, its white snout is fairly long, and it has a white spot above and below each eye.

White raccoons are scarce; pure albinos are even rarer.

The ringtail, *Bassariscus astutus*, lives in the extreme southwestern United States and Mexico. It somewhat resembles a squirrel with a long, bushy, banded tail. Overall it measures about 30 inches, half of which is tail. Its body hair is yellow-gray; the tail is white with blackish rings. Nocturnal, it feeds on insects, lizards, fruits, birds and small mammals. Ringtails, also commonly called ringtail cats, are often seen in pairs and prefer habitats of cliffs, chaparral or rocky ridges near water. In *A Field Guide to the Mammals*, William H. Burt calls the ringtail "a good mouser; probably wholly beneficial."

. . . K.P.H.

RACCOON FACTS

Description: The black robber's mask across the eyes and the banded tail are unmistakable. Body fur is grizzled gray-brown. Generally measures 20 to 30 inches long from nose to tail tip and weighs anywhere from 10 to 35 pounds. Southern raccoons are smaller than those in the North. In the extreme South, like the Florida Keys, adult raccoons may weigh as little as 4 or 5 pounds.

Habitat: Almost everywhere except in very high mountains or desert regions. Farmland, woods, parks, cities, suburbs. Prefers to be near water and large hardwood trees with natural cavities.

Habits: Nocturnal. Sociable with other raccoons to some degree. May den with others during winter.

Den/Nest: Ideal nest and den site is the hollow of a large tree. Will also use manmade structures and, in winter, may den in a woodchuck burrow.

Food: Omnivorous. Eats aquatic animals, fruits, nuts, cultivated grains, birdseed, bird eggs, occasionally a small mammal or bird, garbage.

Voice: Has a number of vocalizations. Whimpers, whines, squeals, cries like a baby as a cub. Screams, churrs and purrs.

Locomotion: A slow, lumbering gait when walking. Can run, with maximum speed barely 15 mph. A good climber.

Life Span: About six years in the wild, perhaps longer. Generally 10 to 14 years in captivity.

FOUR

TIGER SWALLOWTAIL BUTTERFLY
Beauty on Golden Wings

Leaning into the phlox, my face was a mere 6 inches from the tiger swallowtail butterfly. As I watched it probe deep into the center of each pink blossom with its long, black proboscis, I thought back to the butterflies of my youth.

The tiger swallowtail has been an important player in the cast of the backyard wildlife characters I have known since I was seven years old. I used to catch them in a big white butterfly net that my mother had made for me from cheesecloth. Charlie Howe, the publisher of the Tarentum, Pennsylvania *Valley Daily News*, the local newspaper where my dad worked, had a glorious butterfly garden only a few blocks from our home. Much of the year it was resplendent with phlox, daisies, gallardia, primrose and, of course, butterflies. On those warm summer mornings so many years ago, I would sneak into the Howe garden, my net at the ready, and scoop up some flying jewels for my butterfly collection. It was a splendid hobby, and of all the species I caught, the tiger swallowtail was my favorite . . . and still is.

Now, after all these years, I moved an inch closer to the yellow-and-black beauty in my own butterfly garden. I peered into the tiger's black compound eyes. Was it looking back at me as it sucked the nectar from the phlox? How old was it? A day? A week? How much longer would it live? Had it found a mate or mates?

For many people, butterflies are somewhat mystical creatures. They appear in our gardens on a warm day in spring, then come and go through the summer and early autumn—a parade of species changing with the season—and then are gone. While they are with us, they add a marvelous twinkle of beauty to our backyards and to our lives.

In some ways, butterflies are reminiscent of the brilliantly colored birds. Like cardinals, chickadees and goldfinches, butterflies are beautiful to look at, interesting to watch and relatively easy to attract to the garden.

In fact, the same three basic biological needs of songbirds—cover, food and water—are required by butterflies. If a little special consideration is given to the varieties of flowers, shrubs and trees planted, most backyard wildlife habitats automatically have an exciting variety of butterflies present throughout much of the year.

TIGER IS THE PRIZE

Of the many species of common garden butterflies across North America, none is more striking nor admired more by people than the eastern tiger swallowtail. One of the largest and best known, this

With a little consideration to the varieties of flowers, shrubs and trees, most backyards can be attractive to butterflies.

Of the many common North American butterflies, none is more striking or admired by more people than the eastern tiger swallowtail.

goldfinch of the butterfly world is abundant, unafraid and very much at home in a garden environment.

The tiger is a swallow-tailed butterfly, 4 to 6½ inches long. Its wings are bright yellow with a large black V on the inner half of each hind wing. It also has five vertical black bars on each forewing, the innermost two extending to the top of the Vs. When the butterfly is close, like the one I was observing in our garden, some blue can be seen inside the outer margin on the hind wing.

Anyone who has ever touched a butterfly's wings knows that its color easily rubs off. That is because the wings are covered with tiny colored scales. Viewed under a magnifying glass, you can see that they overlap like shingles on a roof. In addition to providing beauty, the scales increase the butterfly's flying efficiency some 15 percent.

Like that of all insects, a butterfly's body is divided into three sections—head, thorax and abdomen. It has six legs, two pairs of wings and two antennae.

THE ORIGINAL BUTTERFLY

The eastern tiger swallowtail gets it scientific name, *Papilio glaucus*, from the Latin *Papilio*, meaning "butterfly," and *glaucus* for the son of the Lycian king Hippolochus, who wore golden armor while fighting at Troy.

The English name "butterfly" also has some interesting roots. One theory is that witches in the form of butterflies stole butter and milk. A more likely explanation is that the name came from the common sulphur butterflies, which are bright butter-yellow. The tiger swallowtail's dazzling yellow also resembles "flying butter."

Though it is most often found in deciduous woodlands, open savannas, and along roads, paths and streams, the tiger swallowtail's strong, sailing flight brings it into gardens and backyards across much of North America. The species ranges from Alaska and the Hudsonian Zone of northern Canada to Florida and Texas.

The tiger belongs to a large family of some 550 swallowtail butterflies found worldwide, believed to be the first family of true butterflies.

The two dozen species of swallowtails found in North America are all strong fliers, and all but two have swallowlike tails.

Tiger swallowtails were among the first butterflies to be illustrated in books. In 1587, a color painting of an eastern tiger swallowtail was shipped from Virginia to England by John White; it was published as a woodcut in 1634. Other drawings and paintings of North American swallowtails were published in the late 1700s and

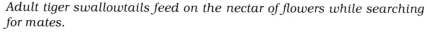

Adult tiger swallowtails feed on the nectar of flowers while searching for mates.

early 1800s to illustrate books on the Lepidoptera—the order of moths, skippers and butterflies.

According to Hamilton A. Tyler, author of *The Swallowtail Butterflies*, members of the tiger group have no Old World relatives. They seem to be of temperate North American origin, evolving through the succession of Pleistocene ice ages.

NECTAR FROM THE GODS

Adult tiger swallowtails feed on the nectar of clover, dame's rocket, thistle, joe-pye weed, bee balm, milkweed and other wildflowers. In gardens, they are partial to phlox, butterfly bush and lilac. In August and early September, it is not unusual for us to see two or three tigers on the tall phlox heads in our southeastern Wisconsin backyard. Many an afternoon tea in our sunroom has been enhanced by the sight of graceful tiger swallowtails gliding around our phlox beds.

The tiger is also a frequent visitor to mud puddles, where it drinks water containing naturally occurring salts. This so-called "puddling" is often a social event at which many, sometimes dozens, of tiger swallowtails gather.

A CHILD OF THE SUN

Butterflies are cold-blooded creatures and therefore depend upon the sun for heat.

Like other butterflies, tigers regulate their body temperatures by positioning themselves to the sun, according to researcher John Edward Rawlins.

In a study at Cornell University, Rawlins found that the black swallowtail, a close relative of the tiger, positioned itself differently according to the ambient temperature. At relatively low temperatures, such as 58 degrees F., the swallowtail positions itself close to the ground with its body raised above its flattened wings for long periods of time. Rawlins believes this helps to elevate the butterfly's body temperature.

At relatively high ambient temperatures, such as 72 degrees F., the swallowtail positions itself with its body lowered in the shade of its wings and remains perched for shorter periods of time. By flying

"Puddling" is a social event at which many tiger swallowtails gather to drink.

more often and perching at greater heights, the butterfly cools itself. With the ambient temperatures between 58 and 72 degrees F., the butterfly is able to regulate its body temperature to between 82 and 89 degrees F.

Rawlins also found that basking in the sun helped the butterflies elevate their body temperatures more than shivering did. He reports that shivering was seen only in disturbed swallowtails under conditions too cool for flight. A minimum body temperature of 75 degrees F. was required before the butterfly could fly, and 82 degrees F. before vigorous flight was possible.

In controlled laboratory experiments, Rawlins found that swallowtails could survive ambient temperatures well over 100 degrees F. if the air was very humid, and below 32 degrees F. for 30-minute intervals.

FOUR STAGES OF DEVELOPMENT

The butterfly is actually the adult or final stage of development of the species. Like the legendary frog that turned into a handsome prince, ugly caterpillars develop into beautiful moths and butterflies.

This metamorphosis doesn't happen instantly. It requires four distinct stages over a period of three to eight weeks, depending on climate, and may even include hibernation.

Stage one of a butterfly's life begins when the adults mate and the female lays eggs. Stage two occurs when the eggs hatch into larvae (caterpillars). After eating and growing, the caterpillar becomes a pupa, which is stage three. As pupae, the insects rest before hatching into adult butterflies.

It has always amazed me how a wormlike creature with a chewing mouth is transformed into a beautiful winged creature with a proboscis for sipping nectar.

Top priority for an adult butterfly is to reproduce. Their movements may appear frivolous as they flit, flutter, flap and glide through trees, and across lawns and land on pink phlox blossoms. But the movements of the tigers are well calculated and with purpose. Their lifetime is limited to only two or three weeks. They must find mates, breed and move on in search of other mates. Sucking nectar and other fluids en route is merely a necessity to keep them going in their search.

In most of its range, both sexes of the eastern tiger swallowtail butterfly look alike. In the South and some parts of the East, at least half of the females are dark brown with marginal spots of yellow, while the males are typically bright yellow and black, with some blue submarginally.

Biologist Lincoln Brower has shown that dark female tiger swallowtails are imitators or mimics of the highly toxic and distasteful pipevine swallowtail butterflies where their ranges overlap in the South. Thus the female tiger enjoys protection from insectivorous birds and other potential enemies.

Male tigers often emerge before females so they can be dry, exercised and physically ready to breed when the females emerge.

A HIT-OR-MISS COURTSHIP

"Girl watching" is the mating game of the male tiger. Most active from about 10:00 A.M. until noon, male tigers generally head for the hills to stake out a territory on any prominent rise in the terrain. They will then patrol their areas as they watch the passing parade. Anything that resembles another tiger is immediately investigated. If it is another male, a fluttering vertical fight will follow. The invader will be sent on his way, undamaged, while the defender will return to his patrolling.

In flatland regions of the country, male tigers will stake out specific flower beds and patrol them in search of females who would be attracted to the same colorful, nectar-filled flowers.

Emerging females, on the other hand, need only flutter to the nearest high point or flowerbed in search of a male.

Initially, sight brings male and female together. But as the two meet in flight and conduct their vertical flutter, recognition of sexual attraction is reinforced by a pheromone (a detection substance) secreted by the female. This stimulates the male to initiate courtship followed by copulation. The pair flutter to a nearby plant, where the male continues to court the female until he can attach his abdomen to hers. Fertilization takes place in about 50 to 70 minutes, according to researcher Steven R. Sims, who studied the sex life of the anise swallowtail, a close relative of the tiger. The sperm are deposited in the female in a neat little capsule.

Once she has been bred, she leaves the hilltop for more private environs in the lowlands. In the evening all the tigers descend to the lowlands, where they roost for the night.

MORE THAN ONE MATING

Females may mate two or three times during their short two to three weeks of life. Males may also perform multiple matings during their somewhat shorter life of about two weeks.

Sims found that male mating ability is low up to 12 hours after emerging from the pupa, that it peaks at age two to eight days, and then gradually declines. After copulation, the male's purpose in life has been achieved.

The tiger swallowtail caterpillar is a green larva with large orange, black-pupiled false eyes.

The butterfly is actually the adult or final stage of development of the species.

The female, however, has more work to do. Once fertilized, she will fly away from her mate in search of a place to lay her eggs. She will make her deposit of the next generation hours or days after mating. As she lays her eggs, the sperm are released from the capsule to fertilize them.

Tiger females are very particular about where they lay their eggs, because the caterpillars which will hatch from them require specific food. When the egg-laden female finds a suitable stand of tulip trees, wild cherry, birch, cottonwood, ash, willow or basswood, she will deposit her 200 to 250 cream-colored, spherical eggs on the leaves, one at a time. She spreads them out over some distance, and over several different trees in the area.

PRESTO! A GREEN CATERPILLAR

In a few days to a week, the eggs will gradually darken. At hatching time they are nearly black. A smooth green caterpillar (larva), looking totally unlike the brilliant-yellow damsel from which it descended, will hatch. Nature has given the tiger caterpillar large, orange fake "eyes" with black pupils located well behind its inconspicuous real head and eyes. When disturbed two bright-orange "horns" shoot out just forward of the eyes, and the air is permeated with an unpleasant odor. Meanwhile, the false head is elevated and sways slowly back and forth in cobra fashion. This deception is intended to scare off predators. The caterpillar also has a series of orange and black stripes traversing its body.

High in a tree, the tiger caterpillar eats its fill and hides in a leaf nest it builds by folding the edges of the leaf over itself and attaching it with a strand of silk from the spinneret in its head.

As the caterpillar grows, it changes skins several times. Each time it molts, its new skin is a little larger than the old.

When it has eaten enough, the caterpillar is ready to enter the third stage of its development—the pupal stage.

A PERIOD OF REST

The caterpillar attaches itself to a twig in a head-up position with a few strands of silk, and then its skin forms a hard shell called a chrysalis. The tiger's chrysalis is mottled light and dark brown to resemble a curled-up dead leaf clinging to the twig.

In the North, female tigers produce two broods a year; in the South, three. The last brood of the summer hibernates over winter as a pupa, emerging as an adult in the spring.

Like an embryo surrounded by an eggshell, the adult butterfly develops inside the chrysalis. At the right moment, the butterfly inside applies pressure, breaks the shell and emerges with wings and body soft, wet and constricted. In a matter of minutes, the insect slowly unfolds as the fluids are pumped through its veins. Finally, when the mature tiger is dry and its wings have hardened, it flies off, resplendent in yellow and black, in search of a mate. Thus, the cycle begins again.

BUTTERFLY IS A SNACK FOR MANY

Many wild creatures relish a meal of butterflies and caterpillars. Songbirds, particularly, are fond of fat green larvae to eat themselves and feed to their young. Furthermore, a great many birds have no

Chipmunk

Cottontail rabbit

Raccoon

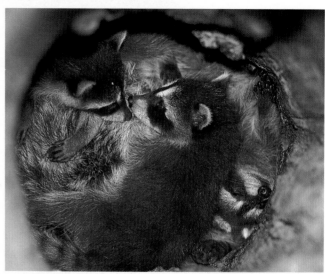

Tiger swallowtail butterfly

Gray squirrel

Striped skunk

White-tailed deer

Flying squirrel

American toad

Woodchuck

White-footed mouse

Box turtle

Opossum

Northern oriole

Evening grosbeak

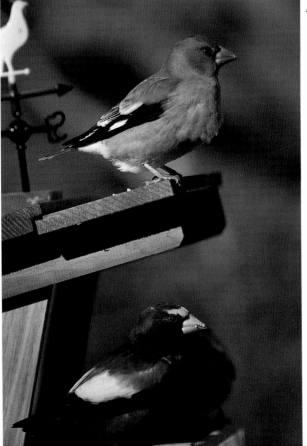

Predators have learned that certain butterfly patterns, such as that of the monarch, are toxic.

qualms about plucking the wings off most butterflies and chomping down the bodies.

Several species of butterflies, however, are toxic. Predators have learned that certain butterfly patterns, such as that of the monarch, are "no-nos," and they leave them alone. One species, the viceroy, which looks very much like the monarch in color and pattern, enjoys protection from predators because its appearance mimics the foul-tasting monarch.

Unfortunately for the tiger swallowtail, only the few females which are dark brown and mimic the pipevine enjoy such protection. Therefore, it must depend on its inconspicuous brown chrysalis and the scary big orange "eyes" on the caterpillar. And it depends on movement—the adult's swift and eluding flight—to save it from predators.

INVITE BUTTERFLIES TO YOUR YARD

Like songbirds, butterflies will respond to specific kinds of food and cover plantings in backyards and gardens.

According to a *National Wildlife* article by Maryanne Newsom-Brighton, a backyard butterfly garden can be as simple as a single nectar-rich plant or as complex as a patch of wildflowers. In other words, with a little extra planning and by providing the right kind of natural habitat, anyone can attract butterflies to his yard.

Butterflies prefer flower blossoms containing lots of nectar and large heads or petals upon which they can sit comfortably while feeding. They are also attracted to certain colors of blossoms. Pink, purple, yellow and white seem to be favored, though nectar is more important to them than color. Therefore, any pink or lavender flower with a large head and lots of nectar should be a butterfly hot spot.

Phlox, alyssum, candytuft, arabis, aubrieta, buddleia, catmint, sedum spectabile, rhododendron, joe-pye weed, milkweed, New England aster, thistle and wild bergamot all fit that description.

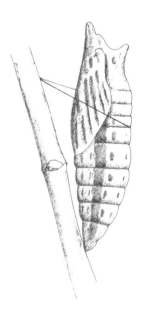

The caterpillar attaches itself to a twig in a head-up position with a few strands of silk, and then its skin forms a chrysalis.

THE OTHER SWALLOWTAILS

The eastern tiger swallowtail has cousins all over North America. Most are as large or larger and equally spectacular.

In the Southeast, gardeners are fond of the somewhat larger and whiter zebra swallowtail, *Eurytides marcellus.* The giant swallowtail, *Papilio cresphontes,* of the East, South and Southwest is about the same size as the tiger, but darker. Still another close relative, the black swallowtail, *Papilio polyxenes,* is colored the reverse of the tiger, being mostly black with a little yellow. It is found throughout nearly all of the continental United States. The smaller pipevine swallowtail, *Battus philenor,* is all black with lovely blue-green hind wings.

. . . G.H.H.

TIGER SWALLOWTAIL FACTS

Description: A bright-yellow-and-black butterfly 4 to 6½ inches long with a large black V on the inner half of each hind wing. It has five black bars on each forewing, the innermost two of which extend to the top of the V. Some blue is evident on the intermargins of the hind wings.

Larva: Smooth green caterpillar with large orange black-pupiled false eyes. Body is traversed with a series of orange and black stripes.

Habitat: Gardens, grasslands, paths, streamsides and deciduous woodlands throughout the eastern two-thirds of North America from the Arctic to the Gulf of Mexico.

Food: Adult—the nectar of flowers, especially pink and lavender, with large heads. Larva—the leaves of wild cherry, tulip trees, basswood, birch, ash and poplar.

Locomotion: Erratic flight.

Life Span: Adult male about two weeks; female slightly longer.

Gray squirrels are among the most interesting and challenging of all backyard wildlife.

FIVE

GRAY SQUIRREL
A Paradox Called Bushytail

"Oh, no! Not another squirrel!"

From my desk, I could see a gray squirrel hanging upside down on a finch feeder, loading up on sunflower seeds. I had just replenished the feeders for the third time because there was a winter storm brewing and I wanted the birds to get their fill before dark.

Years of experience had taught me that it would do no good to try to chase away the squirrel. Gray squirrels just don't scare very easily. After being run off a few times, they simply retreat to the far side of the nearest tree trunk and peek back. (Squirrel hunters call it "sidling" when a squirrel slips around the trunk, always staying on the side away from the gunner.)

A WILDLIFE PARADOX

Gray squirrels are the great paradox of backyard wildlife. On one hand they can be aggressive, destructive, persistent and annoying pests. On the other hand, they can be intelligent, inquisitive, skillful, handsome and among the most interesting wild creatures to observe in the backyard.

Gray squirrels are intelligent, inquisitive, skillful and handsome.

Many people enjoy gray squirrels and spend a great deal of money and time keeping them well fed. Others genuinely hate them and will do anything to rid their yards of the menace.

We Harrisons are somewhere in between. We have a population of a dozen or more healthy gray squirrels at our feeders daily, summer and winter. We tolerate them, and, frankly, we enjoy them (most of the time) as a part of our daily backyard wildlife experience.

At times, we have even become somewhat fond of individual squirrels. There was "Stubby," for example, a fat little female with only half a tail. She was at the feeders from time to time for at least six years. We suspect that she lost part of her tail to a hawk or to mange, a serious disease among gray squirrels when populations grow too large.

Gray squirrels like Stubby are not always as friendly as they appear. My father found that out the hard way when he tried to release one caught in a wire fence. The frightened squirrel bit a 2-inch hole in Dad's hand, which required stitches to stop the bleeding.

Another gray squirrel we watched in our yard for a few years was "Ricky," a male with a ringed tail like that of a raccoon. Because we "knew" him, we tolerated his expensive taste for niger seed.

Ricky was but one of many gluttonous squirrels at our feeders that daily compete with songbirds for the goodies.

The greater conflict, of course, is between gray squirrels and people. This is due to the fact that the natural requirements (cover, food and water) which attract the desirable birds (cardinals and chickadees, for example) to the backyard feeders are the same for gray squirrels.

SOME SQUIRRELLY STORIES

There are endless anecdotes about the adventures people have had while trying to solve gray squirrel problems at bird feeders. At the Schlitz Audubon Center in Milwaukee, Wisconsin, a birdseed study was to be conducted to determine the food preferences of the different species of birds. A dozen Droll Yankee feeders were strung on a wire, each filled with different kind of seed. The study was immediately jeopardized by a horde of local gray squirrels that gobbled up the experimental seed before the birds could get to it. (Incidentally, the squirrels preferred sunflower seed until it was gone; then they ate all the other seeds.)

The Audubon researchers tried stringing beads, coffee cans and 33-rpm records on both ends of the wire to keep the squirrels off the feeders. That failing, they set up large square plastic walls at both ends of the wire, but the squirrels soon learned to leap over the plastic walls, land on the rolling coffee cans and do their balancing act to the feeders. The birdseed experiment was finally abandoned.

Not far away from the Audubon Center, another backyard wildlifer was successful in outwitting the squirrels by placing pie pans at each end of a wire holding his feeders. But the frustrated squirrels finally learned to shake the wire, rocking the feeders until seed was knocked to the ground where they had easy access to it.

One squirrel fan learned a dear lesson from feeding his wild pets in the kitchen of his summer and weekend retreat. He allowed the gray squirrels to follow him inside, where they were invited to eat peanuts and birdseed from the kitchen storage bin. This was a great thrill to the weekender until he returned to his cottage one Friday night to find that the squirrels had not awaited his return. They had chewed their way into the cottage and had helped themselves to the bin of food.

There are endless stories about the adventures people have had while trying to solve the problem of gray squirrels at bird feeders.

Over half of the gray squirrel's total length is made up of its bushy tail.

Even the President of the United States has squirrel problems. One recent spring, the gray squirrels in Lafayette Square, the little park across the street from the White House, ate 2,000 of the President's geraniums, girdled and killed over a half-dozen newly planted trees and seriously injured some 100-year-old oaks.

Nevertheless, gray squirrels can be vastly entertaining creatures to watch. They are exceptional acrobats that seem to delight in a chase through the treetops. That same agility allows them to hang in the craziest positions to get the last seed in a bird feeder.

But there is a great deal more to know about gray squirrels— their social life, sex life and family life and their problems of survival. Much of the information we have about the private lives of gray squirrels was learned only very recently.

THE MEASURE OF A GRAY SQUIRREL

The eastern gray squirrel is a rather large, slim, tree-climbing animal, measuring 18–20 inches in length. About half that length is tail. Adults weigh 1 to 1½ pounds.

Without its silken gray coat, the animal would resemble the ugliest of ratlike rodents. (In some parts of the world, tree squirrels are called "tree rats," as squirrels and rats are both members of the rodent family.)

The back and sides of the gray squirrel's coat are covered with hairs banded with brown and black and tipped in white. Underparts of the squirrel are white, some tinted with brown, tawny or rust. The hairs in its bushy tail are also banded in black and brown with a white tip. All this banding and tipping of hairs gives the gray squirrel its salt-and-pepper appearance.

GRAY IS NOT ALWAYS GRAY

Surprisingly, gray squirrels come in a variety of shades and colors. Some are pure black and some are pure white. Melanism, however, is more common than albinism. In parts of the gray's range, particularly in the northern extremities, one or more youngsters in any litter may be black. This recessive gene manifests itself in shades that vary from jet black to charcoal gray.

In Olney, Illinois, there is a colony of several hundred white squirrels that have become the town's landmark. A city ordinance passed in 1925 protects the albino colony and gives the weaker-eyed white squirrels the right-of-way on city streets. As far as is known, the Olney white squirrels are the only such colony of entirely albino mammals existing anywhere in the world.

Regardless of color, they are all *Sciurus carolinensis*, eastern gray squirrels. *Sciurus* is Greek for "an animal that sits in the shade of its own tail." The species name, *carolinensis*, means that the animal was first named in the Carolinas, though it ranges throughout the United States and southern Canada east of the Rocky Mountains.

MORE THAN A TAIL

The gray squirrel's bushy tail is a multipurpose tool which plays an important role in communication, locomotion and insulation.

In his book *Speaking of Animals*, Alan Devoe describes it. "The tail is to the squirrel what vision is to the hawk, or fleetness to a deer or wariness to a fox. He lives by it."

Gray squirrels come in a variety of shades and colors. Some are pure black.

In Olney, Illinois, there is a colony of several hundred white squirrels that have become the town's trademark.

Of the tail, Ernest Thompson Seton writes, "It is as large as himself, and much more conspicuous. He looks like a huge tail drifting along, pushing a small animal ahead of it."

As a communications tool, the position and movement of the squirrel's tail tell another squirrel a great deal. The tail helps sound the alarm of danger as it is flashed in conjunction with the alarm call. It communicates dominance or submission when two squirrels meet. It is useful in communicating between the sexes when breeding is imminent.

The tail is also a fifth leg, which helps maintain balance when the animal is on treetop excursions, performing tightrope acrobatics and chasing other squirrels. When a squirrel loses its balance and falls out of a tree, the tail acts as a parachute and helps the animal land on all fours. It also helps break the speed of the descent and thus reduces the chances of injury.

Perhaps the most fascinating use of the gray squirrel's tail is as a blanket or muffler to help keep the animal warm in winter, and as a sunshade to cool it in the summer. It even serves as an umbrella in rain or snow.

AT HOME IN THE FOREST

Ideal gray squirrel habitat is deciduous forest, where oaks, beeches and hickories abound. In good years, this kind of woodland provides gray squirrels with all the food they need as well as the proper habitat for raising young. Nuts and large seeds provide a great

The squirrel's tail is a multifaceted tool which plays an important role in communication, locomotion and insulation.

deal of the gray squirrel's diet in the fall and winter. In spring and summer it eats fruits, berries, mushrooms and insects. It will take birdseed anytime it is available.

Gray squirrels also need water at least twice a day. The backyard bird bath, therefore, is an excellent source of water for grays. Our three-tiered recirculating bird pool is visited throughout the year by squirrels almost as much as by birds.

STRANGE MASS MIGRATIONS

One of the strangest historical facts about gray squirrels is their enormous mass migrations. In his marvelous book *Speaking of Animals,* Alan Devoe describes one of these migrations. "On a three-day hunt in Ohio in 1822, the gray squirrels taken were 19,660. Migrating from area to area in the fall of the year, squirrels traveled in gatherings estimated to contain nearly a quarter of a million. There are records, not to be disputed, of bands of squirrels advancing along fronts a hundred miles wide, and requiring five days to pass."

Audubon speaks of seeing the veteran Daniel Boone, then 80 years of age, "barking off squirrels" (see Glossary) on the Kentucky River about 1815. "We moved not a step from the place, for the squirrels were so numerous that it was unnecessary to go after them. . . . procured as many squirrels as we wished."

Another interesting description of this phenomenon was printed in Doutt, Heppenstall and Guilday's *Mammals of Pennsylvania:*

Researcher Vagn Flyger has a number of squirrel boxes in ideal habitat of oak, beeches and hickories.

"When Pennsylvania was first settled, the abundance of gray squirrels amazed our forefathers. The state was rich in forests and the woods teemed with squirrels. At intervals, huge migrations took place and literally thousands of squirrels marched through the woods, ignoring all obstacles in their desire to get from 'here to there.' Thousands drowned in the rivers that blocked their path but until the mysterious urge burned itself out the hordes pushed forward. When the migration was over, the countryside behind was almost devoid of gray squirrels, but no evidence of overpopulation cropped up elsewhere. No one appears to be able to explain just why they travel and where they go."

Lack of food and overpopulation appear to have been the cause of these incredible migrations, which do not occur today. "Probably 1866 was the last of the great treks," according to Seton, "though many lesser movements have been noticed since."

THE SQUIRREL SOCIETY

Squirrels are very social animals. Each has a home range of from less than an acre to over 7 acres, according to a study in Maryland. Other researchers report home ranges from 2 to 20 acres. The shape of the range appears to be long and narrow, according to a Virginia study. The most important fact, however, is that a gray squirrel's range overlaps with the home ranges of other gray squirrels. This is the basis for a squirrel society.

Within the overlapping home ranges, there is a well-defined social hierarchy. When squirrels meet—and they seem to meet often —they recognize each other by sight and smell. This recognition includes the knowledge of which is dominant and which is subordinate. Thus, during most of the year, serious fights are avoided.

After studying gray squirrels for years, biologist D. C. Thompson of the University of Toronto, Canada, reported that the way gray squirrels behave when they meet is important. When approaching another squirrel in a tree, the dominant squirrel crouches slightly, reaches toward the second squirrel with either one or both forepaws and simultaneously springs toward it, pushing off with its hind feet. If both animals are on the ground, the dominant squirrel will run toward its subordinate neighbor with tail held almost flat out behind.

Youngsters must form dominance relationships if they are to establish a home range of their own.

Chasing is also a common form of behavior, with the dominant squirrel doing the chasing. If the subordinate squirrel is overtaken, a short wrestling bout usually results. While wrestling, loud squeals may be heard as the subordinate squirrel is being bitten.

When an unfamiliar gray squirrel enters the home range, the residents of that range will be very aggressive toward the interloper and usually run it off.

Youngsters, when weaned, are considered interlopers by the other residents of the home range area. Therefore, the youngsters must form their own dominance relationships if they are to establish a home range locally. The young of parents that are established residents are more likely to be accepted than the young of strangers. However, about 10 percent of the young are not accepted and must leave the home range of their parents to seek their fortunes elsewhere.

WHEN A SQUIRREL SPEAKS

Gray squirrels are very vocal creatures. Their barking, chattering, screaming, buzzing, mewing, purring and assorted "chucks" are all part of squirrel talk. Further, the flashing and flickering movements of the tail, the stamping of their feet, the ways they walk and the raising of the hairs on their bodies in conjunction with these vocalizations are all important parts of their communication.

The best-known squirrel call is the one that announces the availability of food. Squirrel researchers Frederick S. Barkalow, Jr., and Monica Shorten describe the call as *quac, quac, quac, quaaaaaaaaa . . . quaaaaaaaaee . . . eeeeooooo*. Another call, the one given by females in the breeding season, is so similar to the food call that it is written the same way.

Alarm calls are quite different. The researchers describe the intense, immediate alarm as sounding like *kut, kut, kut,* in loud staccato bursts. When a squirrel is running to safety while giving this alarm, it often pauses and adds *quaa, quaa, auaaee?* as well, according to the researchers. Squirrels hearing it pass on the alarm with a *kut, kut, kut . . . kut quaa . . . quaa . . . quaaa*. This continues until the reason for the alarm has passed, though a quieter chorus of *kuts* and *quaas* with tails flicking and teeth knocking is not unusual.

The alarm call can become so shrill and intense that it sounds

Barking, chattering, screaming, buzzing, mewing, purring and assorted "chucks" are all part of squirrel talk.

almost hysterical. I have been annoyed more than once by this confounded noise while sitting in a woodland trying in vain to be inconspicuous.

Even as nestlings, baby gray squirrels produce a variety of calls. Biologist Robert S. Lishak recorded the calls of four-day-old gray squirrels in their nests on and around the Auburn University campus in Alabama. He describes the sounds of the newborns as "squeaks and lip-smacking sounds." At four weeks of age, the young produced growls, *muk-muk* sounds and screams. Tooth chattering began with the eruption of the incisors at eight weeks, Lishak reported.

A VARIETY OF KEEN SENSES

Any squirrel that uses its voice to communicate as much as the gray squirrel must have a keen sense of hearing.

Its ears are much longer than they often appear to be because they can be furled and laid flat; they can be extended to stand up three-quarters of an inch above the head.

The gray squirrel's large dark eyes, located on the sides of its head, are well suited for life in the trees. It is a diurnal creature, and its night vision is probably no better than ours, but its field of vision is much larger. It can see all around without moving its eyes or head, including forward in binocular vision.

Gray squirrels also have a very keen sense of smell. Their large, mobile nostrils allow them to smell cached food buried under several inches of snow and soil.

TWO LITTERS A YEAR

Gray squirrels can produce two litters a year, one in winter and one in summer. The winter solstice apparently triggers the primary mating season, when a female comes into heat and is pursued in a noisy, energetic chase through the treetops, up and down tree trunks, by two to a dozen or more males.

Squirrel researcher Vagn Flyger of Silver Spring, Maryland, told us that the chase is a vital part of squirrel courtship, and that without it, the female will not ovulate. This is supported by the fact that gray squirrel reproduction in captivity has usually been unsuccessful. Flyger feels this is because there is no opportunity in confinement for the chase.

While breeding, the male secretes a copulatory wax plug which blocks the female's vagina and prohibits further breeding. She becomes solitary and increasingly territorial, chasing away all other squirrels from the den tree she has chosen for her nest. (Males are communal and often den together. Vagn Flyger has found as many as 13 males sleeping in the same nest.)

Squirrels prefer to nest in tree cavities, but if none is available during the summer breeding season, they build bulky leaf nests, which can easily be seen in the tops of trees after the leaves fall in autumn.

Squirrels will also use man-made nesting boxes. Ideally, they are 18 to 28 inches high and about 12 inches square, with a 3-inch hole near the top of the front. They should be placed 20 to 30 feet above

Keen vision, hearing and sense of smell are all vital to the gray squirrel's survival.

Squirrel courtship is a noisy, energetic chase through the treetops.

the ground, according to Vagn Flyger, who has placed hundreds of boxes in his Silver Spring woodland as part of his research. However, both the flicker and wood duck boxes in the front of our Wisconsin home (neither of which meets the ideal specifications) have consistently housed gray squirrel families.

About 44 days after being bred, the female gives birth to a litter of three to five tiny youngsters. They are about 4½ inches long and weigh only ½ ounce. They are helpless and naked, except for little whiskers, and their eyes and ears are sealed. After a week on mother's milk, the babies double in weight.

During their second week, hair starts to push through the skin on their backs. At one month, their ears open and their lower incisors have poked through the gums.

Vagn Flyger told us that "it takes a long time to grow a squirrel." When their eyes finally open at five weeks, they are fully furred, weigh 3 to 4 ounces and measure about 10 inches overall, 4½ inches of which is tail.

Around that time, the babies start to nibble on anything within reach (sometimes each other). That includes any insects that venture close to the nesting cavity. According to a study by Charles M. Nixon in Ohio, insects are an excellent source of protein for juvenile gray squirrels, and they continue to consume them into late spring and summer.

A FLEA MARKET

It is not unusual for a gray squirrel family to utilize two or more nests during the rearing period. Fleas often become so thick and annoying in a squirrel nest that the mother is forced to move her family to another site. She does this by carrying the babies, one at a time, in her mouth. She grasps the baby at the belly with her mouth, allowing the youngster to wrap its feet around her head and neck for the ride to the new den. As she does this, she looks as if she were wearing a rubber collar.

Mother squirrels are attentive and will nurse their youngsters for nine to ten weeks. They will not, however, bring food to them, as bird parents do. When the baby squirrels are between six and seven weeks old, they make their first wobbly forays outside the nest to feed on tender leaves, buds and insects near the nest. By the time they are two months old, they can crack a hickory nut or acorn. The young females, who develop more rapidly than the males, seem to be the most venturesome. The males tend to lag back and return to the nest more often.

These three-week-old youngsters are furred, but their eyes won't open for another two weeks.

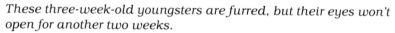

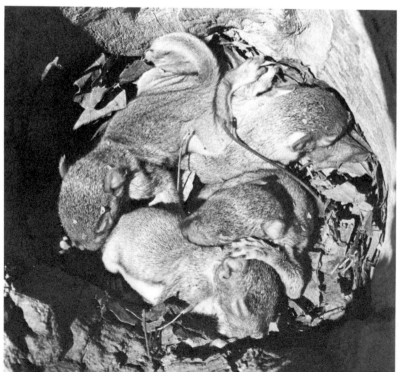

Squirrels prefer to nest in tree cavities.

They'll also use manmade nesting boxes.

ON THEIR OWN WITH SOLID FOOD

At 10 to 12 weeks, baby squirrels are weaned and are fending for themselves on a diet of solid foods. By then their mother has again been bred, following another chase through the treetops by a string of males. Busy preparing herself and a nest for the next litter, which will be born in mid-July to early August, she has no further concern for her last litter.

Meanwhile, the youngsters from her winter litter continue to grow and mature as they devour diverse insects and greens. Researcher Charles Nixon observed three juvenile gray squirrels, about five months old, engaged in what he called "an unusual feeding behavior." From about 35 feet, Nixon watched the squirrels using both their teeth and claws to remove tree bark and eat the insects they exposed. The squirrels expeditiously searched four white oaks, removing pieces of bark as they worked. Soon the ground beneath the trees was covered with bark chips. Nixon later determined that the juvenile squirrels had eaten ants, gall midges, broad-nosed beetles, a centipede, a spider and larvae of oak galls.

Even when fully grown, juvenile squirrels can be identified by the bands in their less luxurious tails.

When they are between six and seven weeks old, baby squirrels make their first wobbly forays outside the nest.

Autumn spurs gray squirrels to hoard like pack rats. They are either eating or stashing nearly everything in sight.

SQUIRRELS GO NUTS

As summer passes into autumn, the diet of gray squirrels, young and adult, changes with the season.

Squirrels begin caching nuts in the ground as soon as there is a surplus. They will not dig them up until food becomes scarce in early winter, and will stop retrieving cached food as soon as there is fresh vegetation available in the spring.

The availability of nuts in the autumn spurs gray squirrels to hoard like pack rats. Suddenly, they are either eating or stashing nearly everything in sight—acorns, hickory nuts, black walnuts, beechnuts, sugar maple seeds, dogwood fruits, wild cherries, black gum fruits, pine seeds and corn.

Fungi also play a principal role in the gray squirrel's diet. According to Vagn Flyger, wild mushrooms, including the deadly amanita, don't supply much energy, but seem to offer necessary nutrients.

Gray squirrels have been known to eat birds' eggs, and will chew on bones and deer antlers for the calcium and phosphorus they contain. Their constantly growing incisors also need gnawing for sharpening.

Though acorns are high on the preference list of gray squirrels, given a choice, they prefer hickory nuts every time, even though they derive less energy from them. Again, the need to sharpen their teeth may be tied to this preference, as hickory nut shells are notoriously hard.

Among acorns, those of the red oak provide the greatest energy for squirrels, while white oak acorns give them the lowest, according to a study in New York by Allen R. Lewis.

Seton, describing the gray squirrel's strong preference for hickory nuts, wrote, "His appetite for hickory nuts amounts almost to a passion. He will pass by all other foods, and brave innumerable dangers for a feast of his favorites. So eager is he for the annual bounty of his mother tree, that he cannot await the decent time of ripening, but cuts them while they are green. He is like an overeager, overgreedy small boy who is too impatient to wait for the thorough baking of his cake, so nibbles and nibbles at the unsatisfactory, unwholesome dough."

SOME FALLACIES ABOUT BURYING NUTS

I like Alan Devoe's description of how a squirrel buries a nut: "Harvest, inspect, eat or store. He takes a nut in his adroit forepaws and turns it carefully over and over, licking it. His sign and seal on it, he scampers off with it a little ways and digs a hole in the earth, perhaps three inches deep. The nut is inserted, point downward, the soil scooped back over it, the disturbed fallen leaves raked and smoothed to look as they did before."

There has been a great deal written about how squirrels bury nuts and how they "remember" where they buried them on lean days many months later.

A big hole was blown in that theory by researchers D. C. Thompson and P. S. Thompson in a study they conducted in Toronto, Ontario. The team proved that squirrels recovered buried nuts by smell, not by memory, and that the squirrel that retrieved the nut usually was not the same one that buried it. A squirrel's sense of smell is so keen that it can sniff out nuts that are under as much as a foot of snow.

NATURE'S TREE PLANTERS

The squirrel's fastidious habit of burying a cache has resulted in much undue credit to them as "nature's little tree planters." Though the Thompsons found some validity in the fact that squirrels inadvertently plant some trees, they also found that of the large numbers of nuts buried by squirrels, 84.6 percent were recovered, leaving a relatively low percentage for possible germination.

In the 1940s, researcher Victor Cahalane of Michigan buried 400 acorns in September and October and marked their locations. By

Squirrels recover buried nuts by smell, not by memory.

December, 60 percent were still where he had buried them. By May, the squirrels had recovered every one of them.

Thus, the bottom line on squirrels' cached food is that it contributes significantly to the winter diet of the general gray squirrel population . . . and possibly to tree reproduction.

NOT HIBERNATORS

Gray squirrels do not hibernate. They remain active throughout the year, except during inordinately cold weather, foraging for a living even on the snowiest days.

On extremely cold days, gray squirrels stay inside their dens, keeping warm and using as little energy as possible. Males will often den together and so share body heat. They may remain in this inactive mode for as much as two weeks, losing some weight, but not enough to starve to death.

MOST LIVE ONE YEAR

Even under the best conditions, the average life span of a gray squirrel in the wild is one year, with only about 25 percent living longer. In captivity, gray squirrels have lived at least 20 years, and biologically they should be able to survive in the wild for at least 12 years, though very few live more than six, according to researcher Frederick S. Barkalow, Jr. (Our Stubby was the exception, perhaps because she lived in a more settled area around our home. We saw her on occasion for at least six years.)

The reason for the high mortality among gray squirrels, as with all animals, is that life in the wild is tough. Predators (such as hawks and owls, black rat snakes, foxes and bobcats), accidents, parasites and diseases (such as mange) take most of the grays. Human hunters usually have little effect on squirrel populations.

I remember interrupting a red-tailed hawk feeding on a gray squirrel while I was strolling through a nearby woodland. The hawk had eaten only the front quarters when I accidentally flushed it.

Gray squirrels have to compete for food with other forest wildlife, such as ruffed grouse, deer, wild turkey, black bear, jays and other species of squirrels. This competitive coexistence is the ecological structure of the forest community. It is significant that a great deal of that rivalry is nonexistent at backyard feeders.

SOME SQUIRREL-PROOF SOLUTIONS

It is possible to squirrel-proof your bird feeders, but only under certain conditions. A squirrel baffle, for example, can be mounted on the post under a bird feeder. The best baffle is shaped like a stovepipe. It allows the squirrel to climb up the post and into the closed pipe, but no farther. The pipe must be long enough to keep the squirrel from climbing over it. Some homemade baffles are constructed of a wastebasket mounted upside down on the post. Another good baffle is made from the aluminum duct used for venting clothes dryers.

These baffles work *if* the feeder is placed far enough away from nearby trees, wires, buildings, or any object from which the squirrels can jump onto the top of the feeder.

Some people outwit squirrels by greasing the feeder post. This works only if the grease is present in sufficient quantities to keep the squirrels on the skids.

There is a commercially made sliding post which operates on the same principle as the greased post. The squirrel climbs the spring-loaded double-lined post toward the feeder until the weight of its own body lowers it back to the ground. (I am told that with some practice, squirrels can learn to outwit that one, too.)

To keep squirrels off hanging feeders, more challenging solutions are required. Suspending the feeders from piano wire, like those at the Schlitz Audubon Center, is doomed to failure. In spite of coffee cans and beads to make the high-wire act more enjoyable to watch, even the least agile gray squirrel can master that setup in a couple of days.

Gray squirrels remain active throughout the year, foraging for a living even on the snowiest days.

It's possible to squirrel-proof bird feeders, but only under certain conditions.

The most effective way to keep squirrels off hanging feeders is to use squirrel-proof feeders. Some are covered with wire mesh large enough to allow most birds to enter the feeding chamber to eat, but small enough to exclude gray squirrels.

Anne Gromme of Briggsville, Wisconsin, designed a simple bird feeder which is squirrel-proof, sparrow-proof, starling-proof and weatherproof. She uses plastic lids at both ends of a coffee can. A half-moon is cut out of the top third of both lids, and the feeder is suspended horizontally from a string attached to the can and hung from an S hook.

Many of the plastic tube-shaped finch feeders now have metal covers around the seed holes. This prevents the squirrels from chewing the plastic to enlarge the holes.

THE ULTIMATE SOLUTION

If you really can't stand another day of seeing squirrels at your bird feeders, the only solution may be to get rid of them. One birder expressed it this way: "We consider the squirrels weeds in our garden and we merely remove them."

There are several ways to remove squirrels from your yard, but none is easy. If you don't have a large population of gray squirrels, the best method is to live-trap them and transport them far enough away from the feeding station to guarantee that they will not return.

The eastern fox squirrel is a larger tree squirrel with tawny-tipped hairs.

That means taking them at least 10 miles away; otherwise the squirrel may be back home before you are. Live-trapping without injuring the animal can be accomplished with Havahart traps manufactured by the Havahart Trap Company, Lititz, PA 17543. These traps are sold in most hardware stores and are available in different sizes to accommodate animals ranging from mice to raccoons.

Another effective removal method, though perhaps distasteful to many, is to consider them a product of the land and harvest them during legal hunting seasons.

We believe that the best solution is to tolerate them. This may be accomplished by designating one tray feeder as the official squirrel platter. There you must offer a good supply of corn, sunflower seeds and peanuts or whatever is required to keep the squirrels off the birds' feeders. In the long run, this is the ultimate solution.

THE OTHER SQUIRRELS

There are two other species of gray squirrels in the United States. The western gray squirrel, *Sciurus griseus*, is larger than the eastern and lacks the brownish tinge, but otherwise is almost identical in appearance and life-style. It lives only along the West Coast and in the mountains of the West Coast. The other look-alike is the Arizona gray, *Sciurus arizonensis*, which lives in Arizona, New Mexico and Mexico.

Other tree squirrels in the same genus are *Sciurus aberti,* the tassel-eared squirrel, which is found in the four states of the Southwest. *Sciurus niger,* the eastern fox squirrel, whose range overlaps the eastern gray, is a larger tree squirrel with yellow to orange belly and tawny-tipped hairs in the tail. In the Southeast, fox squirrels may have mixtures of yellow, white and black on their bodies and heads.

The other North American tree squirrels which are not *Sciurus* are the red and chickaree squirrels of the genus *Tamiasciurus* and the flying squirrels of the genus *Glaucomys.* Both are smaller and are quite different in appearance. Though they prefer different habitats from the grays, both overlap a great deal of the gray's range, and compete with it in many backyards across the United States. Flying squirrels have their own chapter in this book.

. . . G.H.H.

GRAY SQUIRREL FACTS

Description: Large gray tree squirrel, 8 to 11 inches long, with bushy tail measuring an additional 8 to 10 inches.

Habitat: Hardwood forests with nut trees, river bottoms and wooded suburban backyards.

Habits: Active during daylight, all year. Seldom far from trees.

Nest/Den: Tree cavities in winter; sometimes leaf nests in summer; will use man-made nesting boxes.

Food: Tree nuts, fruits, seeds, buds, flowers, mushrooms, birdseed.

Voice: Noisy. A rapid *kut, kut, kut* means danger; less rapid means danger has passed. Also chatters, screams, barks, mews, purrs and whines.

Locomotion: Scurries along the ground at about 12 mph; travels with ease from tree to tree by jumping.

Life Span: Average one year in the wild, but potential for at least 12½ years. Over 20 years in captivity.

George and his pet skunk, circa 1940.

SIX

STRIPED SKUNK
Master of Chemical Warfare

As a child, I was enchanted with Flower, the gentle skunk friend of Bambi. Flower—beautiful, sweet and amiable—was content to float through life sniffing blossoms. Flower was interested in neither provoking nor being provoked. Because of Flower, I believed skunks were delightful creatures. I still do.

Skunks truly are gentle, handsome animals, with personalities closer to Flower's than most realize. They demand nothing more than to be allowed to plod along minding their own business.

Several years ago, while George and I were at Cape May, New Jersey, for the autumn hawk migration, we had an encounter with a skunk that emphasized this.

We had dragged ourselves out of bed well before dawn. As a matter of fact, it was still pitch-black outside as we crawled into our cold car, sipped coffee from paper cups and headed for one of Cape May's birding hot spots—the corner of Sunset Boulevard and Cape Avenue in a residential neighborhood.

When we arrived, it was still at least 30 minutes before dawn. We wanted to be in place when the hawks and songbirds took to the skies with the first light.

Standing quietly in the chill, damp stillness, we became aware of a muffled scratching sound on the other side of the crepe myrtle. It stopped; we heard some faint gruntings, then there was silence. We froze . . . those were the sounds of a skunk digging for grubs or insects and contentedly grunting while devouring the morsels. We remained quiet, hoping it would trundle off in the opposite direction to crawl into its den and sleep for the rest of the day. Alas, it was not to be.

We saw the distinctive black-and-white pattern come around the end of the stand of crepe myrtle. It headed directly for us just as the first bird, a rose-breasted grosbeak, zipped past. As the large skunk rounded the turn, we saw another, smaller black-and-white form follow, then another and another. This mother skunk was leading a single-file procession of six half-grown skunk kits.

George's hair tingled on the back of his neck. I wondered if I'd be able to stay motionless, or at least keep my movements slow and controlled. With that first rose-breasted grosbeak, the birds started to race past us. Catbirds, vireos, orioles, tanagers, kingbirds and warblers were flying with such speed and frenzy in their determination to get south that birds nearly flew into us on two occasions. This was what we had come to Cape May to experience. What a thrill! Hard as it was, we had to stifle our excitement until the skunk family was beyond us. As long as we made no sudden movements, we knew we had nothing to fear from the skunks.

At a distance of about 12 feet, the mother skunk stopped, looked George straight in the eye and then veered off to the right. Her kits

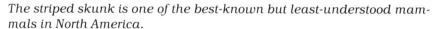

The striped skunk is one of the best-known but least-understood mammals in North America.

followed, we would have sworn, in her exact footsteps. The caravan made a short bypass, then, when beyond us, continued on their way, a rippling stream of black-and-white fur.

Meanwhile, the hawks had started to move through the area. Had we yielded to the impulse of pointing and shouting, "There's a Cooper's hawk!" or "Duck! Wow . . . that Swainson's thrush almost hit you!" we might have ended up "skunked."

BEST-KNOWN MAMMAL

The skunk is perhaps our most common and most widespread carnivore. William H. Burt, author of *A Field Guide to the Mammals*, believes it is the best-known mammal in his book.

The skunk is about the size of a house cat. It averages 24 to 30 inches from nose to tail tip, and usually weighs from 4 to 10 pounds, though some excessively fat individuals have tipped the scales at 14 pounds. Males are about 10 percent heavier than females and as much as 8 percent longer.

The skunk seems to be built in wedges. The pointed snout, broadening to the ears, forms a wedge-shaped head. The body, narrow at the shoulders and wider at the hips, is a larger wedge. The white markings on its otherwise glossy black coat start as a thin line from the nose and flare to a wedge shape at the back of the head before splitting into a V of two lines running down the back and onto the tail.

The tail itself, responsible for one-third of the animal's total length, is magnificent. It is a glorious brush, rivaling the bushy plume worn by the fox. The skunk characteristically carries its resplendent appendage slightly aloft as it toddles along in search of the next grasshopper.

POWERFUL WEAPONRY

Under the base of the tail is the skunk's notorious ammunition. Two grape-sized glands, one on each side of the anus, hold a pungent musk that the skunk fires at will by contracting the surrounding gizzardlike mass of muscle. On each scent gland is a small nipplelike outlet which can be aimed with lightning speed and excruciating accuracy. For this reason, *Mephitis,* a Latin word meaning "noxious gas," was thought to be an appropriate scientific name for the skunk. The striped skunk is *Mephitis mephitis.*

Other members of the weasel family are also provided with musk glands under the tail. However, "they reach their glorious perfection in the skunk and furnish it with a wonderfully effective weapon of defense," Victorian naturalist Ernest Thompson Seton declared.

The musk is a clear amber oily fluid. Its odor is indescribable, yet familiar to everyone. Seton, groping to put the sensation into words, suggested, "Imagine a mixture of strong ammonia, essence of garlic, burning sulphur, a volume of sewer gas, a vitriol spray, a dash of perfume musk, all mixed together and intensified a thousand times. This will give a faint, far, washed-out idea of skunk musk." In *Mammals of Pennsylvania,* Doutt, Heppenstall and Guilday colorfully describe the smell as "a combination of steel mill, rotten egg and the tiger house at the zoo, as penetrating as ammonia."

Actually, Seton, like a number of other naturalists, had a special fondness for skunks even as a youth, and thought that the diluted smell of skunk was quite pleasant. In his boyhood days in Toronto, a popular legend held that a skunk had wandered into town years before and had been killed on the southeast corner of Pembroke and Gerard streets. "On damp nights, it was easy to perceive the undying odor of the creature on that very spot. As a child, I used to linger near," he admitted, "and go out of my way to enjoy it, for it made a marvelous appeal to my wildwood instincts," he said, adding, "The fact that it later turned out to be nothing but a leakage of sewer gas can never rob me of the joyous thrills and memories of those long-gone times."

A RAISED TAIL MEANS TROUBLE

A skunk prefers to avoid situations that call for such actions, but if provoked may reluctantly let loose a barrage. Interestingly, fighting skunks don't seem to use musk on each other. "Evidently it would be wasted in such a combat," Seton speculated. "It would be like two ducks splashing each other."

When in the presence of a skunk, it is wise to remain as quiet as possible. Sudden movements frighten skunks and are interpreted as threatening.

Typically, the skunk goes through three warnings before resort-

The third and final warning is when the tail tip is raised. The body is bent into a U so that the skunk can see where it's spraying.

ing to a squirt. First, it stamps it front feet, then it raises its tail with tip downward. "The third, final and dreadful warning is when the tip rises up and spreads out," Seton warned. "That white flag nailed to the mast does not mean 'surrender,' but clear the deck for action. Then you look out! Stand perfectly still! Make no sudden move; it may not yet be too late!" he counseled. "The skunk, especially if an experienced old fellow, may change its mind, haul down the fighting signal, mast and all, forgive you, and go quietly away."

When the skunk discharges its "noxious vapor," it's facing its victim with both front and back end for the instant it takes to fire the jet. Its body is bent into a U, so that the sharpshooter sees where it's firing, and inevitably does so with incredible accuracy up to at least 14 feet, usually targeting the eyes.

Skunk musk is *n* butyl mercaptan, a highly sulfurous compound which can cause temporary blindness. When sprayed in the dark, it is luminous, like "an attenuated stream of phosphoric light," according to Audubon's colleague the Rev. John Bachman.

POTENT POTION

If necessary, a skunk can fire five or six consecutive rounds. Each dose may be a mere fraction of a teaspoon, but it's more than adequate.

By lifting the tail high, arched over its back, the skunk manages to avoid getting any on itself. A prudent maneuver, because even one drop has astounding power.

George came to this realization the hard way. When he was a budding naturalist of 12 or 13, he decided to try his hand at taxidermy. Riding along a country road on the way home one night, young George spotted a dead skunk in the road. Oh, boy! A specimen for practicing his taxidermy! "Dad! Stop the car . . . please! I *need* that skunk!" George's father, being a professional naturalist, and his mother, being moderately indulgent, agreed to retrieve the roadkill only after determining that the skunk had died without spraying. Its scent glands were intact. Nevertheless, George's younger sister, Gretchen, instinctively made a sour face and uttered, "Phew! Skunk!" as the body was placed in the rear of the station wagon.

The next day, forbidden from working on the skunk in the basement as he had been allowed to do with his other specimens, George set up an operating table in the garage. This was a big event, and he made certain he had an impressive audience before beginning. All the neighborhood kids crowded around as George poised the scalpel for the first incision. Gingerly he inserted the sharp tip to make a cut down the abdomen, taking great care—he was certain—to avoid the fully loaded musk glands. Not enough care. In frozen horror he watched as one small amber droplet rose from the gland.

Kids scattered every which way, banging into each other in their panic to get out of there. George, his face close to his work, got a concentrated dose. Even though it was only one tiny drop, he gets a little nauseated even to this day when he smells skunk. That's why the hairs on the back of his neck stood on end when the skunk family sauntered past us in Cape May.

In his youth, George found a dead skunk on the road and took it home for a disastrous taxidermy episode.

A BRACING TONIC

John Burroughs, like Seton, quite enjoyed skunk scent. "It approaches the sublime, and makes the nose tingle," he wrote. "It is tonic and bracing, and I can readily believe has rare medicinal qualities."

As a matter of fact, skunk musk was occasionally used to treat asthma in the 1800s. "We were once requested by a venerable clergyman, an esteemed friend, who had for many years been a martyr to violent paroxysms of asthma, to procure for him the glands of a skunk," John Bachman reported. "According to the prescription of his medical advisor, [they] were kept tightly corked in a smelling bottle which was applied to his nose when the symptoms of the disease appeared," he explained.

"For some time he believed that he had found a specific remedy for his distressing complaint; we were, however, subsequently informed that, having uncorked the bottle on one occasion while in the pulpit during service, his congregation, finding the smell too powerful for their olfactories, made a hasty retreat, leaving him nearly alone in the church."

DOGS NEVER LEARN

Dogs are among the most common victims of the skunk's chemical warfare. They delight in chasing small animals, and skunks are no exception. Hunters watch in terror as their setter, pointer or retriever takes off after a black-and-white plume. No amount of whistling or yelling seems to dissuade a dog hot on the trail of a skunk. If the hunter tries to keep up with the dog, hoping to pull him away, he may accomplish nothing more than sharing in the consequences of having frightened the skunk. When they return home, he and the dog may be firmly told to find other sleeping quarters.

There are dozens of reputed antidotes for neutralizing skunk odor on clothing and dogs. Tomato juice, liberally poured over the dog and rubbed in vigorously, is one of the favored remedies. "It works very well," claims Dick DeGraaf, a principal research wildlife biologist with the U.S. Forest Service. "It takes about six large cans for one dog and one boy!" Some swear that the lotion from a home permanent deactivates the scent.

One of the problems is that when your dog gets a blast of skunk juice, it's nigh impossible to endure being around it long enough to administer any of these concoctions. George's prescription is lemon-scented after-shave lotion. It doesn't neutralize the odor, but it has such a strong, permeating smell of its own that it does a respectable job of masking the skunk scent.

OTHER SURVIVAL GEAR IS WEAK

Equipped with such a powerful weapon, the skunk has not had a need to develop keen senses for survival. Its hearing is weak, its vision is poor, its sense of taste and smell are mediocre.

It cannot climb, nor has it developed speed on the ground. It plods along in a slow, ambling, flat-footed gait at about 1 mph. Skunks have no reason to hurry, but they can speed up to what might be called a "trot" of 3 or 4 mph. At top speed, they may attain 6 mph. No wonder dogs overtake them so easily.

Skunks are capable of making a number of sounds, but are ordinarily silent. Occasionally one will churr or scold if it is disturbed or make low grunting sounds while feeding, probably to express pleasure. It can also growl, snarl, twitter like a bird and squeal. When it's angry or issuing a defiant warning, it stamps its front feet.

It plods along in a slow, ambling, flat-footed gait.

A GENTLE CREATURE

In spite of its potent defense mechanism—or perhaps because of it—the skunk is an easygoing, shy yet utterly self-confident creature. It seems to be aware of its powers, displaying an unconcerned attitude toward man and beast. "Most human beings armed with a weapon of offense so powerful that none dare face it are apt to become arrogant with power, to degenerate into bullies and tyrants," Seton philosophized. "The skunk is as far as possible from such behavior. His natural sweetness of character has never deserted him in his 'glut of power.' He continues the gentlest and least aggressive creature in the woods."

Alan Devoe and his wife Mary had an extraordinary experience on a March evening with a wild skunk. In *Our Animal Neighbors,* they tell of coming across a young skunk that had gotten its head lodged tightly in a tin can. Devoe motioned Mary away to a safe distance, then knelt in the snow and talked softly to the skunk.

"Instantly he turned, lifted his tin-hooded head and came to me," Devoe related. "The tin can bumped against my knees. The skunk backed a few inches and stood still. Now or never."

Devoe stroked the skunk's back. It stood perfectly still, except for the forepaws, which were treading the snow. Devoe looked at the tail. "It stayed down in a skunklike signal of trustfulness and unalarm."

With his right hand on the skunk's shoulders, Devoe grabbed the can with his left hand and pulled. "I could feel the small body stiffen and see the skunk brace his paws, but he seemed to cooperate by pulling backward. But the can did not come off." The top of the can

In spite of its potent defense mechanism, the skunk is an easygoing, shy creature.

The striped skunk favors semi-open areas, fields and pastures, and rural and urban neighborhoods.

had not been removed when it was opened; it had been pushed inward. "It had been easy for a hungry and inquisitive *Mephitis* to poke his head inside," Devoe said, "but the angle of the can held him prisoner, almost like a trap treadle, when he tried to withdraw."

Devoe turned the can slowly, worked it forward and backward, turned it, pulled, pushed and maneuvered. The little skunk was nearly lifted off the ground with all the tugging and twisting, but the tail stayed down. "There was just a squirming, struggling little black-and-white body, cold and snow-wet, making mighty efforts of cooperation," he said.

Then the can came loose, taking a tuft of skunk fur with it. The freed skunk blinked in the sudden light. "I got to my feet very slowly, so as not to startle our newly freed *Mephitis*," Devoe continued. He and Mary stood quietly in the snow. "The skunk sniffed at our feet, peered up at our faces, and then for several magical moments brushed his body and tail back and forth across our legs as a happy cat will do. He gave himself the luxury of a prodigious shaking, spattering us with snow and water," Devoe recalled. "He sniffed a time or two at the snow-filled air, turned from us, and went rocking off in his crooked-tracked way."

HAPPINESS IS A WOODCHUCK HOLE

The striped skunk is found throughout the United States except in desert regions, north into southern Canada and south as far as Guatemala.

It favors semi-open areas, grassy and weedy fields and pastures, brushland, open prairie, dumps, and both rural and urban neighborhoods. Usually it is no farther than 2 miles from water.

Wherever it lives, top priority for a skunk is to establish a den site. Occasionally a skunk will use its strong, straight claws to dig an underground burrow. More often it uses an abandoned woodchuck burrow or foxhole.

It may also set up housekeeping in or under a stump or fallen log, in a rock pile or a woodpile, or under a back porch, a shed or other outbuilding.

THAW BRINGS COURTSHIP

With the February thaws, the skunk breeding season begins, and it continues through the end of March, peaking in mid-February. Fresh out of his winter den, the male skunk pads about looking for a mate. If he finds a receptive female, he'll probably move in with her. He might stay only a short time, impregnating her and then moving on, or he may settle in for a lengthy visit.

If he is still there around April 1, the female "notifies her spouse in some way that he is no longer needed around the house," according to Seton, "and in her judgment, it is an excellent time for him to try a change of life and air."

With her domicile to herself, the female relines the den with dried grasses and leaves which she pushes into the burrow.

IT'S A FAMILY

In late April or May, 63 days after mating, the female delivers up to ten mouse-sized, ½-ounce kits, but a typical litter numbers five or six. The newborns' nearly naked pinkish skin, thinly covered with short, fine fur, plainly shows the black-and-white color pattern. With eyes and ears sealed, the infants are blind and deaf at birth.

The female, an attentive mother, cuddles her young and nurses them by sprawling over them or lying on her side.

At two weeks, the youngsters are fully furred and weigh about 4½ ounces. At 17 to 21 days, their eyes and ears open. By the age of

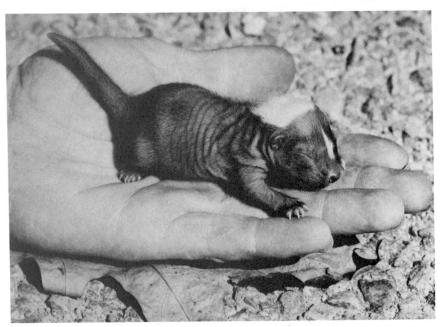

Even newborn skunks plainly show the black-and-white color pattern they'll have throughout life.

one month, the kits weigh about 12 ounces and make their first shaky attempts at walking on their short legs. Nevertheless, they will try to stamp their feet and scold if angered, and even try to assume the stance for spraying, though their little musk glands are barely the size of peas.

INDIAN-FILE OUTINGS

In late June or July, when the kits are about seven weeks old and have grown to as much as 1½ pounds, they make their first excursions outside the den with their mother, following her in single-file formation.

Small duplicates of their mother, the little skunks poke their noses into every hole, rock and tussock that their mother does. They solemnly mimic her every move, copy every sniff, every pawing.

Within their home range of ½ mile or so, the skunk family may travel as much as 4 miles or more as they trundle haphazardly about, inspecting all possible food sources.

A BALANCED SKUNK DIET

Strictly speaking, skunks are omnivorous, but about 70 percent of their diet is animal matter, primarily insects like grasshoppers and crickets and small mammals like mice, voles and shrews. So effectively did they destroy the hop grub in New York State that legislation to protect skunks was passed at the insistence of the hop growers.

With their strong claws, skunks can dig out mouse nests, tear apart rotted logs to get at insect larvae and expose turtle eggs buried in sand or soil.

Sometimes they'll eat the eggs of a ground-nesting bird, but their inability to climb keeps them out of higher nests.

They often make little diggings 3 to 4 inches in diameter and an inch or so deep in their search for insects, grubs and earthworms. At times, there are dozens of these little scrapings in our backyard, but they are usually inconspicuous and never harm our lawn.

The other 30 percent of what skunks eat is vegetable matter, mostly in the form of fruits, grasses, berries and buds.

FAMILY LIFE CONTINUES

Shortly after they begin accompanying their mother, the youngsters are weaned, but they stay with her into autumn and, particularly in the North, may share her winter den. Others leave the family unit in early fall.

A mother skunk moves her young in the same way that a mother cat transports her kittens.

Skunks often make little diggings in their search for insects, grubs and earthworms.

By October or November, the young of the year could be mistaken for adults except for their smaller size. They, like their elders, have put on a thick padding of fat in preparation for winter.

And, like the adults, they have the same confident aura as they putter along on their nightly outings. They have little to fear, because nearly every potential predator seems to know instinctively that skunks are best left unmolested.

Unfortunately, skunks haven't learned that their unique and powerful defense mechanism doesn't intimidate cars. More skunks probably perish on the highways than from any other cause.

LUXURIANT FUR

Many skunks are trapped each year, too, whether deliberately or coincidentally, in traps set for furbearers. Its luxurious, rich, glossy black fur is of good quality. Traditionally, the less white in the fur, the more desirable the pelt, so furriers established a four-point system of grading skunk furs, with four points assigned to those that are nearly all black.

Years ago, when tremendous numbers of skunk pelts were shipped to U. S. markets and to London for the European market, the fur was often labeled "Alaskan sable" or "black marten." Seton used to chuckle about this. "If, as often happened at some fashionable 'tea,' where women wore their furs, it came on one of those close, warm, or damp days so well known in London, the stimulating effect on every skunk skin present often resulted in an atmosphere which caused an early break-up of the gathering," he claimed.

FEW NATURAL PREDATORS

Among wild things, great horned owls are probably the greatest threat to skunks. They have no sense of smell, so no matter how often they get sprayed, it doesn't deter them from going after another skunk dinner. It's not unusual to find a great horned owl's nest surrounded by skunk skeletons. While searching for a great horned owl's nest a few years ago near Aldo Leopold's famous shack in central Wisconsin, George and I followed our noses. We sniffed out the nest by the telltale musk of the skunks the adult birds had brought to the nest for their young.

In some years, rabies can be epidemic in a local skunk population. Rabid skunks, obviously, are to be avoided. Normally skunks are timid. One with rabies may become aggressive and show no fear of people. Any wild skunk that deliberately approaches is suspect, and local authorities should be notified.

Even with few natural enemies, skunks probably average lives of only two to three years in the wild. In captivity, they might live to ten years.

PUTTING THE SKUNK TO USE

Overall, the skunk benefits man, not only with its long fur, but also by destroying enormous quantities of insect pests and vermin.

Chippewa Indians rendered skunk fat and used it as a rubbing oil. They also drank it as a purgative for worms.

According to early accounts, Indians and trappers used to eat skunk meat regularly. C. Hart Merriam claimed that the meat is white, tender and sweet and is delicious eating, much like chicken, but more delicate. "Being, happily, free from any of that 'squeamishness' which Audubon and Bachman lament as preventing them from tasting the meat of this animal, I am able to speak on this point from ample personal experience," he bragged.

Strange as it may seem, skunk musk can be used as a fixative in perfume. When I was a child, my father sometimes teased me when I put on cologne in my attempt to be more grown-up. With a sly grin he'd ask, "Is that Chanel No. 5 or Hot Afternoon on a Skunk Farm No. 1?" He may have been closer to the truth than either of us realized.

ENTERTAINING PETS

When surgically deodorized, skunks make good pets. We've known several that belonged to various friends.

Deodorized skunks have sometimes been available in pet shops. Today, at least in our part of the country, it is illegal to have a skunk in captivity because of a rabies outbreak that began a couple of years ago.

It seems that all the prominent early naturalists kept skunks, at least for a while. Merriam had a clever young male that he named "Meph," a shortened form of the skunk's scientific name. "After supper, I commonly took a walk, and he always followed close at my heels," Merriam wrote in 1884. "If I chanced to walk too fast for him, he would scold and stamp with his forefeet, and if I persisted in keeping too far ahead, would turn about, disgusted, and make off in an opposite direction; but if I stopped and called him, he would hurry along at a short ambling pace and soon overtake me.

"He was particularly fond of ladies," Merriam noted. "He would invariably leave me to follow any lady that chanced to come near."

Merriam and Meph used to walk to a large meadow abounding in grasshoppers. "Here Meph would revel in his favorite food, and it was rich sport to watch his maneuvers. When a grasshopper jumped, he jumped, and I have seen him with as many as three in his mouth and two under his forepaws at one time. He would eat so many that his over-distended little belly actually dragged upon the ground," Merriam declared.

Enjoy your skunks, but enjoy them from a distance.

Meph's nest was in a box at the foot of the stairs. Before he was strong enough to climb out by himself, he would stand on his hind legs with his paws resting on the edge of the box when he heard Merriam coming and would beg to be carried upstairs. If Merriam passed without appearing to notice, Meph chippered and scolded, stamping his feet vehemently. He enjoyed being in Merriam's office, and as soon as he was strong enough, he'd climb up the stairs on his own.

DENNING FOR WINTER

Before severely cold weather arrives, skunks prepare their winter dens. In the South, they remain active most of the winter. In the North, they sleep through the coldest months, but they do not hibernate.

Communal winter denning is common among skunks, but there is typically no more than one adult male per den. Many use the same winter den they occupied the previous year.

M. E. Sunquist, studying winter activity of striped skunks in east-central Minnesota, found that in his study area, the average date on which adult skunks began using their winter dens was November 17, compared to November 24 for juveniles. All study animals, according to Sunquist, were in their dens before permanent snow cover.

The skunks in Sunquist's study area commonly stayed in their winter dens until March. By the time they emerged in spring, they had lost half of their late-autumn weight.

HUNGRY SKUNKS

Offspring from the previous year that stayed with their mother through the winter leave the family unit at this point. Both the yearlings and the adults are ready to breed as soon as they come out of their winter dens.

This is also the time of year when food is scarcest for skunks. On mild nights in February and March, it's not surprising for us to notice a hungry skunk slowly waddling across our patio, earnestly looking for a tidbit—perhaps a few crumbs of suet that fell to the ground when the birds were working on the tree-trunk suet feeder.

As with raccoons, some people enjoy feeding skunks at their back door. Often, the offerings are readily accepted if the skunks are allowed to dine in peace. Chances are that they will supplement their diet with the proffered food for two weeks or so, then fall back on their natural diet as spring progresses.

ENJOY THEM AT A DISTANCE

There's no need to fear having skunks in your backyard if you leave them alone. Skunk squirt aside, keep in mind that familiarity is not a good idea because of the unpredictability or rabies occurrences. Enjoy your skunks, but enjoy them from a distance.

I don't know of anyone who enjoyed his skunks more than Seton did. "What bright, happy hours they were that I spent with glossy, playful broods. I have no language to explain the little thrills of pleasure they begot," he said. "I used to tell myself that it was the sense of success in the raising of fur. Of course, I knew it was not; for the idea of killing one of the exquisite creatures for its fur was horrible. My pleasure was that of the naturalist, happy in seeing and in being among abundance of joyous animal life. I used to make excuses to go and sit among them," he confessed. "Not by day only, but at night, oftentimes after a dinner or theater party or a late opera, if it were moonlight, I would slip away from the family to linger in silent happiness among the merry, fur-clad throng. In a sort of wordless joy I would sit, as they gambolled about or climbed into my lap or pawed my foot in scratchy invitation to a romp. I would hold one up and gaze into his beady eyes, and try to reach and read his little soul," he reminisced. "Those were good times indeed!"

AMERICA'S OWN

Skunks are endemic to America, with the striped skunk being the most widely distributed.

The eastern spotted skunk, *Spilogale putorius*, and western spotted skunk, *Spilogale gracilis*, are handsome animals, smaller than the striped skunk. On jet-black fur they sport a white spot on the forehead and under each ear; four broken white stripes along the neck, back and side; and a white tip on the tail. They are able to climb trees, but are normally terrestrial. Spotted skunks are found throughout the United States and into Mexico with the exception of the Great Lakes states, New England, the mid-Atlantic Coast, Montana and parts of Washington, Wyoming and North Dakota. Don't look for the same warning signs given by the striped skunk. When spraying, a spotted does a handstand on its forefeet and sprays directly over its head.

The hognose skunk, *Conepatus mesoleucus*, is distinguished by a naked piglike snout and an entirely white back and tail. The rest of the body is black. It is an animal of the extreme Southwest, occurring only in New Mexico, Arizona, southern Colorado and Texas, as well as in Mexico.

The hooded skunk, *Mephitis macroura*, barely ranges into the United States. It occurs in Mexico and in southern New Mexico and Arizona. The hooded skunk may be nearly all black with two white side stripes, or its back may be mainly white, including the very long tail. It gets its name from the hair on its neck which spreads into a ruff.

. . . K.P.H.

SKUNK FACTS

Description: A cat-sized animal. Its long black fur is marked with white starting in a thin line between the eyes, broadening to a wedge at the back of the head, then separating into two lines running down the back and usually onto the tail. Tail is a long, lush plume. Feet are rather short, snout is pointed, ears are small.

Habitat: Open woodlands; brushy fields, urban, suburban and rural backyards.

Habits: Nocturnal. In North, sleeps during most of winter. May be gregarious in winter den. When threatened, sprays a pungent musk from scent glands under the tail.

Den/Nest: An abandoned woodchuck, fox or badger hole, if available, or in a rock pile, woodpile, hollow stump or log, or under an outbuilding.

Food: Omnivorous. Diet is 70 percent animal (primarily insects and some small mammals) and nearly 30 percent vegetable (buds, berries, fruits and grasses).

Voice: Usually silent, but sometimes softly grunts when eating. Can snarl, growl, churr, twitter like a bird.

Locomotion: A slow, ambling gait of about 1 mph. Top speed when running is about 6 mph.

Life Span: Probably one or two years in the wild. Has lived to ten years in captivity.

The most common hoofed wild animal in North America, the white-tailed deer is also the most widely distributed and most popular of the big game species.

SEVEN

WHITE-TAILED DEER
Backyard Monarch

Taking careful aim at one of the white-tailed deer eating quietly at the bird feeding station, the rifleman sitting in the open kitchen window squeezed the trigger.

"Bang."

Wheeling around, the deer raised their white-flag tails and galloped off into the woods.

One of the deer ran slowly. When it reached the woods, it stopped and collapsed.

What is happening? Is this a hunting story? How unsportsmanlike to shoot a deer at a feeding station from a kitchen window!

Yes and no. The deer was shot not with a bullet but with a tranquilizer dart. The location was the Schlitz Audubon Center, Milwaukee, Wisconsin. The reason for immobilizing the deer was to transfer the animal and some 60 more of its kind out of town to reduce the deer herd on the nature center's grounds and the surrounding suburb.

Strange as it may seem, there are so many deer in some residential areas across North America that they have become a serious problem. The people living around the Schlitz Audubon Center, for example, have been happily feeding deer in their backyards for years.

145

In many towns and cities, deer have become so numerous that they're causing problems, like these at Milwaukee's Schlitz Audubon Center.

But the herd has grown to a point where the animals are eating everything and anything green, including hedges, shrubs and flowers. They are also eating too much of the natural vegetation on the adjacent nature center. Something had to be done.

A deer removal program was the answer. Eighteen deer were immobilized and transported to a Department of Natural Resources tract some 30 miles away at a cost of $100 per animal. Some 15 to 20 deer were to be removed each year after that. It is an expensive, but acceptable, solution to the problem.

Milwaukee is not the only metropolitan area where there are too many deer. In suburbs all over North America, deer have become so common that they are regular visitors to many backyard feeding stations, day and night. In some cities and towns, deer have been unwanted visitors to downtown shopping areas, where they have crashed through store windows, bashed into car fenders and galloped through parks and playgrounds.

The fact is, there are more deer in North America today than ever before—at least 20 million white-tailed and mule deer, and the number continues to grow.

ONCE AN ENDANGERED SPECIES

It doesn't take much searching through state conservation department records to find reports of excitement over the mere sighting of a single deer. If the term "endangered species" had been invented in the early 1900s, the white-tailed deer would have been on the list. It was extinct in much of its range.

Fortunately, all of that has changed today, thanks to timber harvests which created optimum food conditions and to sportsman-financed management programs.

A whitetail's preferred historical habitat is forest edges, swamp borders and woodland openings. Today, white-tailed deer are found in many backyards that offer that same kind of natural habitat. People all over the country now sit in their breakfast rooms munching on Bran Chex while deer munch on tulips just outside their windows. A pest? Yes, but what a gorgeous pest! Even the most jaded of backyard wildlifers has to admit that there is a very special thrill in seeing a wild white-tailed deer.

We love to see the deer around our home in southeastern Wisconsin. From time to time we give them treats in the form of ear corn, apples and alfalfa. The gray squirrels usually get the corn before the deer, but they love the apples and alfalfa, especially when we have had a heavy snow.

THE CLASSIC BIG GAME MAMMAL

The most common hoofed wild animal in North America, the white-tailed deer, *Odocoileus virginianus*, is also the most widely distributed and most popular of the big game species. It ranges from coast to coast and from central Canada, throughout the United States and Mexico, to northern South America.

A large animal which stands 38 to 40 inches high at the shoulder, the white-tailed deer is 4 to 6 feet long from nose to the base of its 7- to 11-inch tail. The average weight of an adult is 125 to 175 pounds, with bucks (males) weighing more than does (females). A number have weighed over 400 pounds, and two over 500 pounds.

In summer, the whitetail's coat is reddish above the white underparts, and grayish-brown to nearly blue in winter.

While in velvet, the buck's antlers are covered with a soft, tender membrane which can easily be injured.

A whitetail's preferred historical habitat is forest edges, swamp borders and woodland openings. Today, they are found in many backyards offering the same kind of natural habitat.

Buck deer annually grow antlers. Each antler of a typical white-tail "rack" has a main beam and several points. Bucks grow antlers through the summer, then shed them following the breeding season in early winter.

THE WHITETAIL'S TAIL

The whitetail's principal field mark is its white tail, which is white only on the underside. On top it is tan and sometimes partially black, fringed with white. The deer often raises its tail, showing a white flag, as it flees in alarm.

One of the most recent studies of the purpose of the whitetail's white flag was conducted by David H. Hirth and Dale R. McCullough at the University of Michigan, Ann Arbor. They concluded that the flag "serves as a socially cohesive signal that helps keep individuals in groups for antipredator benefits." In other words, the purpose of the signal is to alert other deer in the group of danger so that the group can flee in a unit.

VOICE COMMUNICATIONS

The whitetail's white-flag alarm is often accompanied by a snort or "blow."

"I have heard both bucks and does blatting with a rather toneless sound much like that an old ram sheep would make," stated fellow nature writer Leonard Lee Rue III in *The World of the White-tailed Deer.* "The voice of the fawn has been compared to the bleating of a lamb. As the fawn grows older, its voice sounds more calf-like but lacks the volume and depth of sound," Rue concluded.

Deer will also communicate threats by stomping their front feet. While photographing a large "tame" buck in a research enclosure at Penn State University a few years ago, I watched a threatening whitetail buck communicate to me his aggressiveness by stamping his front feet. He warned me to retreat by raising one foot, hesitating with it in midair, and then stamping it onto the ground several

The whitetail's white-flag alarm is often accompanied by a snort or "blow."

times. This signal was accompanied by the waving of his massive set of antlers, which were poised for attack.

I got the buck's message, but before I could vacate his woodland turf, the normally friendly buck charged and injured a deer researcher who had accompanied me. Had my companion not been carrying a lead pipe, which he had to use, we both might have been killed.

That same week in neighboring Ohio, a Division of Wildlife photographer was killed in an almost identical situation.

OTHER SENSES ALSO KEEN

A deer's sense of smell is very important to its survival. Not only can it detect its fellows, it can also smell the enemy upwind. Anyone with deer around his home knows that they will approach at rather close range if you are downwind from them. Once they get a whiff of human scent, the tails go up and they are gone.

In addition, deer are sensitive to the natural warning sounds of the woodlands, such as a blue jay's alarm call, the scolding of a gray squirrel or the barking of a neighborhood dog which might suggest approaching danger. They can detect any alien sound and are on guard until it is identified. Their ears are like turrets, ever swinging this way and that, searching the wind for danger.

Combined with acute eyesight (they see shades of gray, not color, and will spot any movement at an incredible distance), the white-tailed deer is a living fortress of sensory devices well designed to protect itself against both human and natural predators.

MAN REPLACED NATURAL PREDATORS

Before humans flooded the continent, white-tailed deer were mostly the prey of mountain lions and wolves. Except for free-running dogs, which now inflict the highest predator toll, man exercises the greatest control over deer numbers.

Scientists estimate that natural reproduction accounts for a 20 to 30 percent increase in deer numbers at the end of each summer. This dramatic leap in the population is then reduced by natural mortality, hunting and accidents. In suburban areas, like the community surrounding the Schlitz Audubon Center, these controls are absent.

Buck deer experience dramatic body changes in late summer as their antlers grow to maturity and their necks swell.

In addition to the legal and illegal harvest of some 4 to 5 million deer annually in the United States, another 200,000 are accidentally killed each year on the nation's highways (30,000 on Pennsylvania highways alone).

Deer also suffer losses from winter starvation, a variety of diseases (most do not affect humans) and blood-sucking insect pests.

The bottom line is that the average buck deer lives only one and a half to two and a half years. Does average a year or two longer, though one was known to live more than 19 years in captivity.

The most accurate method for aging deer in the field is by examining tooth replacement and wear. Having worked on deer-hunting check stations for both the Virginia and Pennsylvania Game Commissions, I learned how to age deer using this method. After I had looked at a couple of hundred sets of teeth, it was easy to determine the animal's age.

In a laboratory, biologists can remove an incisor tooth, slice it, stain it and count the annual layers of cementum for positive aging.

AUTUMN BREEDING SEASON

Unlike most kinds of wildlife, which breed in the spring, deer are sexually active from October through December as frost and snow signal the start of winter in the northern hemisphere.

Male deer experience dramatic body changes in late summer as their antlers grow to maturity and their necks swell. They are in a

The soft velvet coverings on their antlers are rubbed off on the trunks of small trees and brush, creating scrapes on the vegetation which are called "buck rubs."

breeding condition known as the "rut." The soft velvet coverings on their antlers are rubbed off on the trunks of small trees and brush, creating scrapes on the vegetation which are called "buck rubs."

In the process of rubbing off the velvet and polishing its antlers, a buck also rubs forehead glands on the brush, depositing scent that will inform other bucks and does of his presence.

Additional scent glands are located on the legs and feet of all deer regardless of sex and age. They include the tarsal glands on the inside of each hind leg at the hock joint; metatarsal glands on the outside of the hind leg halfway between the hoof and the hock; and interdigital glands located deep between the toes of each foot. Any hunter who has field-dressed a deer is well aware of the potency of these scent glands.

THE POINT OF ANTLERS

A buck uses his antlers to fight other bucks for the right to breed with does in his area. Combat between buck deer is not idle play. It is serious, and sometimes deadly, business as the opponents try to gore each other with sharp antler points and kick each other with cutting hooves. (Backyard wildlifers must steer clear of bucks in the rut.) There is usually a great deal of pushing and shoving, and when one of the combatants tires or is injured, it will retreat in defeat.

It is quite rare for two bucks to lock antlers in such a way that they cannot be separated. The result of this predicament is death for both. Seton tells a story about this with an unusual twist. "I remember reading an account of a hunter finding two bucks thus locked—

one dead, the other near death. He was a humane man, so went home for a saw and cut the living one free. The moment it felt at liberty it turned its feeble remaining strength on its deliverer, and he had much ado to save his own life before he could regain his rifle and lay the ingrate low."

Audubon and Bachman tell of *three* bucks whose antlers were interlocked.

Buck deer are neither monogamous nor do they maintain a harem. They are animals of opportunity and breed whatever does are available to them. Their home range may be an oval of 2 or 3 square miles, depending on topography, food and habitat. During the breeding season, their range may expand to 10 to 12 square miles.

Deer in some of the prime whitetail habitat of the upper Midwest, East and South live in concentrations of more than 30 deer per square mile.

MOTHER OF THE HERD

Meanwhile, the doe deer, objects of all this strenuous activity of the bucks, become fertile with the onset of estrus in October and are receptive to the males for only a 24- to 30-hour period. That condition is repeated at 28-day intervals at least three times, or until the doe is

Bucks sometimes lock antlers in such a way that they cannot be separated. Often, the result of this predicament is death for both.

bred, whichever happens sooner. Once bred, birth of fawns will follow in about 205 days.

In doe fawns which are in good physical condition, estrus usually occurs for the first time when they are seven to eight months of age. Fawns of these young does are born late the following spring. Males do not breed until their second autumn when they are a year and a half old.

Contrary to popular belief, healthy does are capable of producing fawns throughout their lifetimes. One doe in Wisconsin lived 19½ years and produced her last fawn in her 18th year. They may produce single fawns, twins and, in rare cases, triplets. Yearling does usually produce a single fawn, but after that, twins are the rule in a healthy herd. The severity of the winter and the competition for available food are the most critical influences upon fawn production. If a doe is under pressure to survive a hard winter, she will produce only one fawn or none at all, despite the two embryos she may carry. If it is a matter of her life and death, the doe's body will resorb one or both of her fetuses before they come to term.

THE LONG, HARD WINTER

When all the fighting and breeding is over, bucks shed their antlers and settle down to a more sociable and quieter life. They band

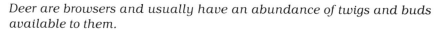

Deer are browsers and usually have an abundance of twigs and buds available to them.

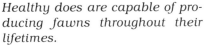

Healthy does are capable of producing fawns throughout their lifetimes.

Deer, like all hoofed animals, get bogged down in snow.

together with does and fawns to face the winter in herds which can number into the hundreds in deep snow country.

An old doe is usually the leader of the whitetail social order. Her group of fawns, bucks and younger does follow her as she works her way through the home range of the herd, following well-worn trails, eating, drinking, and bedding down.

THE TROUBLE WITH SNOW

Deer are browsers and usually have an abundance of twigs and buds available to them from deciduous plants and coniferous growth. They also graze on grasses, herbs and mushrooms. At times, they will grub for roots. In backyards, they will eat corn, apples, alfalfa and the tender greens in the vegetable garden.

They are ruminants, digesting their food by chewing a cud and processing it through a four-compartment stomach, like a cow.

But in deep snow, all hoofed animals have problems acquiring food. Not only is much of their food hidden, but the animals get bogged down in the snow while searching for it. Their weight per square foot is just not distributed well enough to support them on top of snow, unlike a man on snowshoes or skis, for example.

Their long legs, which act as stilts, are about the only advantage deer have in deep snow, but even stilts are of no use in snow which is belly-deep. In this condition, deer are vulnerable to predators as they flounder and become immobilized.

For that reason, during periods of deep snow, deer will restrict their movements to specific protected areas called "yards." Typically, large numbers of deer stay together in yards where the snow is packed down and where there may be insufficient browse to support the herd over the period of snowy weather. Feeding trails leading in and out of the yards are well trampled and offer firmer footing for the members of the herd.

Researchers at the University of New Hampshire have found that deer have the amazing ability to slow their metabolism during periods of deepest winter, thus requiring less food.

We also know that the hair of a deer is uniquely constructed to insulate the animal against the coldest temperatures. Individual hairs are coarse, brittle and crinkled. Under a magnifying glass you can see that they are hollow and contain air. When the animal stands its hair on end, as a bird fluffs its feathers, the insulation qualities are accentuated.

YARDS CAN BE DEATHTRAPS

How well deer get along in their yards depends a great deal on how long they are confined to them. A deer yard can become a death-trap when overbrowsed. Prolonged periods of deep snow and severely cold weather eventually lead to starvation and a lower reproductive rate the following spring.

Because they are smaller and have shorter necks than the adults, fawns born the previous spring are at a tragic disadvantage in yards where the browse line is out of their reach.

Having visited a few deer yards where gaunt-looking whitetails were too weak to stand, I can tell you that winter weather can be a grim reaper. Wildlife biologist A. Starker Leopold, the late son of Aldo Leopold, estimated that the nation's winter deer loss to starvation can run as high as 2 million animals.

It is during severe winter weather conditions that deer often visit backyards, searching for corn and other grains set out for birds and smaller mammals.

During periods of deep snow, deer will restrict their movements to specific sheltered areas called "yards."

SPRING AND NEW LIFE

No matter how severe, no winter lasts forever. The pressure is finally off when deer glean those first tender green shoots growing on the southern exposures where snow has melted. The hard season is finally over.

In most parts of the country where deer are common, the fawning season peaks during late May and early June. No one has witnessed more fawns being born in the wild than Leonard Lee Rue III: "As the moment of birth arrives, a sense of urgency dominates the doe, and she becomes very restless. With heaving flanks and opened mouth, she lies down, then arises only to lie down again. Her body strains and her movements aid in her labor. In a normal birth, the forefeet of the fawn appear first, followed quickly by its head. The fawn, during birth, is in a position suggesting someone about to dive into the water. The entire birth may require only ten minutes, or it may consume an hour."

Whitetail fawns weigh 5 to 6 pounds at birth. Like the young of all hoofed animals, they are born in an advanced state of development.

"Immediately after birth," Rue continues, "the doe gives her fawns a thorough washing by licking them with her rough tongue. She seems aglow with pride and as anxious as a human mother to be certain that her baby is perfect, as she goes over each fawn inch by

Yearling does usually produce a single fawn, but after that, twins are the rule in a healthy herd.

A fawn's tawny-red coat is broken with about 300 white spots, causing it to blend perfectly with its surroundings.

inch. The young are by this time attempting to stand. The washing by the mother deer is usually so vigorous that it knocks the unsteady little fawns off their wobbly feet," he relates.

Because newborn fawns are so unsteady on their feet, they have difficulty in reaching the doe's udder. Therefore, the doe often lies down for the first nursing, Rue tells us.

Like little lambs, the fawns punch and pull at the doe's udder and wag their little tails in their enthusiasm. They can drink as much as 8 ounces of milk in less than a minute, Rue states, though the mother may have trouble providing that much each time the fawns nurse.

Rue goes on to say that as soon as the fawns are able to walk, which is when they are about an hour old, the doe will lead them away from the place of birth before they are discovered by a predator.

If a natural predator, such as a bobcat or fox, should discover a fawn, the doe will usually defend her youngster by attacking the predator with her dangerously accurate sharp hooves. An enraged doe is real trouble for most predators that have designs on a fawn.

SPOTTED FOR PROTECTION

During its first week of life, the fawn is left alone, except when nursing, to lie motionless in the protective cover of brush and leaves. Its tawny-red coat is broken with about 300 white spots, causing it to blend perfectly with its surroundings. Furthermore, it has little or no scent to attract predators during its first few days of life.

Meanwhile, the mother is nearby, but not close enough to draw attention to her youngster. It is during this critical period that people intervene, find the "abandoned" fawn and remove it to the local Humane Society or wildlife rehabilitation center. Thus, thousands of fawns are orphaned each spring. Nearly all the fawns I have seen and photographed were orphaned just that way.

If left alone, the fawn will be cared for by the mother, which returns to her youngster eight to ten times in a 24-hour period.

Leonard Lee Rue tells about the time when he was photographing a red-eyed vireo from a blind near his home in New Jersey. He had entered his blind before daylight and was unaware of a fawn that lay some 75 feet in front. When a doe appeared, he watched as she approached her fawn, which Rue then saw for the first time: "After briefly touching noses, the fawn started to nurse and when it had finished, the doe started to leave. The fawn followed. Again and again the doe turned and swept the fawn back by pushing against it with her head. The fawn persisted until at last the doe raised her forefoot, placed it on the fawn's back, and pressed it to the ground. At this, the fawn lay still, and the doe passed out of my sight."

A few years ago in Maine, Kit and I were sure that a doe we were watching had a fawn hidden somewhere nearby. We even heard the fawn bleating, but we could not find it.

RICHNESS BRINGS GROWTH

The milk of a doe deer is very rich, with about twice the solids of cow's milk and nearly three times as much fat and protein, according to John Madson in *The White-tailed Deer*.

The fawn's inactive period of hiding lasts for only a week or so. By the time they have reached the ripe old age of two to three weeks, they are nipping on grass tips and tender twigs at their mother's heels. Madson states that fawns nurse heavily for nearly two months, but may be completely weaned by the time they are three and a half months old.

NEW ANTLER GROWTH

While does are tending to fawns, bucks are growing a new set of antlers. Unlike sheep and cattle, which grow permanent horns, deer

Deer beds are easy to identify by the matted-down grass or brush.

Starting as buds in May, the buck's antlers reach the velvet stage by summer and are polished by the fall breeding season.

replace their antlers every year, starting in April or May as buds on the buck's brow. At first, the antlers are covered with a soft and tender membrane, charged with blood vessels which are easily injured and bleed. Bucks protect their sensitive and growing antlers, because injury to them will cause a deformity in the finished antler.

The size of the antler is not determined by age, but by nutrition. Poorly fed bucks produce spikes or forkhorns after their first year. Well-fed bucks, on the other hand, grow an average of two to four points on each antler.

In September, when the antlers have reached their maximum growth for the year, male hormones cause a cut-off of blood and the antler dries up and hardens. The velvet is rubbed off against saplings and brush. The monarch is again ready to do battle.

SOME RARITIES

About one doe out of every 4,000 will have enough male hormones in her to grow antlers.

Another abnormality is the piebald deer, a fairly common mutant which is partially white. The rarer all-white albinos have pink eyes and white hooves because of a total lack of pigment. There are also black deer that have an overproduction of the pigment melanin. Rue believes that white deer not only have more difficulty hiding from danger, they also have problems hearing. Most of these freak deer are ignored by other deer, which appear to recognize their abnormality and vulnerability.

MAKE DANGEROUS PETS

White-tailed fawns have been described as the most beautiful and graceful creatures in the wild. For that reason, and because they are so vulnerable at birth, many wind up in captivity. Those responsible for this believe that they can raise this beautiful animal to be a friend and companion for years to come.

But like most wild animals, the older a deer grows, the less adaptable it is to captivity. Adult deer, especially bucks, are dangerous wild animals that will never behave like a domestic dog or cat. They just are not suited to a restrained life with man.

There are numerous stories, like the one told earlier, of a captive buck in the rut turning dangerous overnight and attacking its keeper, sometimes inflicting a severe injury or death.

It is far better to admire deer for their beauty and grace in the wild. Today, that means as close as the backyard. As long as wild deer remain wild and afraid of people, they will never be dangerous. It is the wild deer that is raised in captivity and has no fear of humans that can be dangerous during the rut. A farmer friend had just such an experience when a young buck, apparently raised in captivity, attacked him while he was walking through his woodlot one sunny autumn afternoon. The deer pinned my friend to the ground, tore his clothing and scraped and bruised his body before he could escape.

White-tailed fawns have been described as the most beautiful and graceful creatures in the wild.

There are numerous stories of captive bucks in the rut that turned dangerous overnight.

Like white-tailed deer, mule deer are so common in parts of their range that they frequent backyards with suitable habitat.

HOW TO ATTRACT DEER TO THE BACKYARD

If there are already deer in your general area, they can probably be coaxed into your yard with alfalfa, ear corn or apples during the winter, and a soybean patch in the summer.

Deer are suckers for salt. A salt block of the kind that farmers feed to their cattle and many of us use in our soft-water tanks will perform magic in attracting deer to the backyard or to wherever you would like to see them. Rue says that it would be more beneficial to the deer if mineral blocks were used instead of salt blocks.

Deer are also big drinkers, and therefore a pond or pool, which will serve virtually all the other backyard wildlife in this book, may also be tempting to a family of whitetails.

THE OTHER AMERICAN DEER

The two species of deer living on the North American continent are descendants of a common ancestor of the Pliocene period 10 million years ago.

The Virginia white-tailed deer, *Odocoileus virginianus*, has seventeen subspecies. They include the large northern woodland whitetail of the Northeast and upper Midwest; the Coues or Arizona whitetail (larger ears) of the Southwest and Mexico; the Columbian whitetail of the Northwest; and the diminutive Florida Key deer.

The other species, the mule and black-tailed deer, *Odocoileus hemionus*, lives in the western half of North America from Alaska and Canada to Mexico. It is slightly larger than the whitetail and is distinguished by its big, mulelike ears and the black-tipped or all black top side of its tail. Mule deer antlers differ from whitetail antlers. Each antler forks into two nearly equal branches rather than growing from a single central stem.

Like white-tailed deer, mule deer are so common in parts of their range that they frequent backyards with suitable habitat.

. . . G.H.H.

WHITE-TAILED DEER FACTS

Description: A large hoofed animal, 38 to 40 inches high at the shoulder, 4 to 6 feet long. Hair is reddish in summer, brownish-gray to nearly blue in winter. Well-fed males annually grow antlers with 4 to 8 or more points.

Habitat: Forest edges, swamp borders, woodland openings and many suburban backyards.

Habits: Mainly crepuscular and gregarious, forming small family groups led by old does. Flashes white undertail when alarmed.

Bed: Deer curls into spot on ground within 2- to 3-square-mile oval-shaped home range.

Food: Browses on a variety of woody deciduous plants and some coniferous growth; grazes on grasses, herbs and mushrooms. Also grubs for roots.

Voice: Snorts when alarmed. Also blats and whistles. Fawns bleat.

Locomotion: Runs at normal speeds of 35 mph, maximum of 45 mph; gallops, trots, leaps and jumps to 8.5 feet vertically and 27 feet horizontally. Good swimmer.

Life span: Few bucks live more than 1½ years due to hunter harvest, and does a year or two longer, though normal life in the wild can be 11 to 12 years. Captive deer have survived to nearly 20.

EIGHT

FLYING SQUIRREL
Pixie of the Night

There were at least a dozen out there. I could hear them, but, like fairies of the night, the flying squirrels were experts at keeping themselves invisible unless they wanted to be seen. Each time one of them "flew," I heard the unmistakable sound of its claws quietly slapping against a tree trunk as it landed. The sound was as soft as a large plop of rain hitting a basswood leaf.

A bullfrog began croaking his lazy song from the lakeshore 50 feet away. Hawk moths and an assortment of smaller insects danced in the glow of the patio lights.

Out of the corner of my eye I thought I saw some movement. Yes, there was another one. The nocturnal squirrels were getting closer, moving in to the sunflower and niger seed feeders just as they had every night that summer. Beyond the illumination of our patio lights, I could see a small, flat form gliding from one tree to another, like a large leaf sailing to the forest floor. Within minutes, one flying squirrel delighted me with its arrival on the tree trunk a yard in front of me. The soft slapping sound against the bark drew my eyes to the spot just in time to see a 5-inch-long bundle of reddish-brown fur, with a straight, flat tail and eyes that seemed much too large, scamper around to the opposite side of the trunk. The next time I

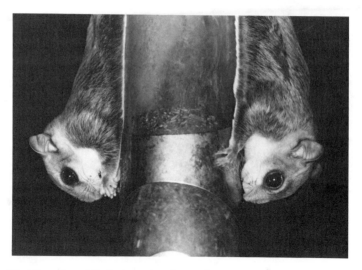

The family of flying squirrels at the authors' feeders were so tame they allowed themselves to be photographed at close range.

saw the little pixie, it was eyeing me from a vantage point 8 feet higher. Small, circular hairless patches around the nipples on her belly were barely visible, but when I saw them I was certain this was the female that had recently introduced her brood of four bright-eyed youngsters to our feeders.

Like all of her species, this flying squirrel was quite tame, and it didn't take long for her to work her way back down to the tubular sunflower seed feeder where she had been enjoying the offerings every night for the last few months. She had brought her four small offspring with her for a while. Now they were coming in on their own, differing from her only in their slightly smaller size and the somewhat pewter cast in their fawn-colored coats.

By the time she was settled in on this particular night, two of the youngsters had claimed portholes on the niger-seed feeder, hanging upside down and using their dainty tongues to lick the seeds out one at a time.

As the mother became more comfortable with my presence, I slowly lifted my hand toward her, gently reached out . . . and touched her! Her coat was incredibly soft and velvety. I was thrilled! She, on the other hand, gave no reaction whatsoever; she just continued to munch sunflower seeds. Her enormous, liquid black eyes, finely rimmed with black fur, gave her an appearance that was irresistibly sweet yet puckish at the same time. Somehow, flying squirrels always look as though they're ready for mischief at any moment.

Flying squirrels are easy to identify because they are the only nocturnal squirrels.

Meanwhile, the two young ones were keeping up a dialogue of their own in soft twitters. Beyond the patio, I saw and heard other flying squirrels gliding among the trees, chasing, scolding, and probably waiting for an opening at the seed feeders or suet holder.

SQUIRRELS OF THE NIGHT

The name given to flying squirrels by science, *Glaucomys volans*, translates to "gray mouse that flies." Like the common name, this, too, is a misnomer. Flying squirrels would more properly be called gliding squirrels, for they glide, or volplane; they do not fly.

They are easy to identify, because flying squirrels are the only nocturnal squirrels. They are much smaller than gray squirrels, about 9 inches long from nose to tail tip, and weigh little more than 2 ounces.

Flying squirrels are reddish-brown or grayish-brown on their backs and pure white on their underparts, with a blackish border running down their sides separating the two colors.

Besides their oversize eyes, another distinguishing characteristic of flying squirrels is their patagia. The patagium is a loose flap of skin, fully furred on both sides, that extends from the wrist of the foreleg to the ankle of the hind leg. This flap makes it possible for flying squirrels to become airborne. It is the animal's gliding apparatus and parachute. It also makes the squirrel look as if its skin were a few sizes too large when it is sitting.

The tail, not bushy like a gray squirrel's, is flat, with the hairs lying outward from the center, like vanes on a feather. It may act a bit like a rudder, but is most important for braking. The tail is thrown upward just before landing to reduce speed, lessening the impact.

Whether in the company of others of their kind or alone, the flying squirrels we've known have been chatterboxes. They have a repertoire of vocalizations, including a scold which is similar to, but much softer than, the scold of red or gray squirrels. On a few restless summer nights, unable to sleep, we've heard them scolding. It's almost a squeak, but not quite that sharp. Naturalist Ernest Thompson Seton described it as "not unlike the complaint of a red-eyed vireo whose nest is threatened." I don't think it's as slurred, or whining, as that. In any case, it leaves no doubt that the caller is dismayed or annoyed.

A quiet, continuous chippering seems to be given when they're a little nervous, as when they're approaching the bird feeders for the first visit of the night and see me grinning at them from the other side of the sunflower seeds. When they're relaxed and eating contentedly, the chippering softens to a twitter.

HAPPIEST IN HARDWOODS

Southern flying squirrels are the most abundant squirrels in woodlands and backyards throughout the eastern United States, from southern Florida and eastern Texas to the Great Lakes and Nova Scotia. Because they are nocturnal, few people are familiar with them; most are surprised to learn that they are fairly regular backyard visitors.

They live among deciduous trees, with mature hickories, maples, beeches and oaks among their favorite tree species. They are found in urban as well as rural and wilderness areas, as long as there are trees to provide them with food, lodging and mobility and there is a source of water nearby.

Flying squirrels are not strongly territorial. Females might defend a small home range, but males generally don't. Their home ranges usually are about ½ acre, but on occasion an animal may wander across 4 acres, or rarely, even up to a mile in a night.

Though seen only at night, flying squirrels may be the most abundant squirrels in the woodlands and backyards of North America.

Abandoned woodpecker holes or natural tree cavities are the most common homes of flying squirrels.

AT HOME IN THE TREES

Abandoned woodpecker holes or natural tree cavities 20 to 30 feet above the ground are the most common homes of flying squirrels. There they sleep during the day, huddle for warmth in winter and raise their young. They will also build nests in birdhouses, under roofs and in attics. Each individual has several nests, usually with entrances that measure 1.6 to 2 inches in diameter, according to P. G. Dolan and D. C. Carter.

Sometimes a flying squirrel uses an outdoor nest, constructed much like the leaf nests of gray squirrels. In fact, they've been known to use deserted gray squirrel leaf nests.

Because of their nocturnal habits, flying squirrels haven't been studied as much as their diurnal cousins, so there is still a great deal to be learned about their courtship and family life. Much of what we know is the result of observing captive flying squirrels.

They usually breed twice a year, once in early spring (probably February or March), and again in midsummer, after the young from the first litter are out of the nest.

About 40 days after the mother is bred, the tiny squirrels are born in a den lined with shredded bark and other soft material, including fur which the mother plucks from her breast.

A typical litter size is three, but there may be from two to six offspring. At birth, they don't have the appealing beauty of their parents. Weighing about ⅕ ounce, with eyes and ears tightly sealed and toes fused, the 2-inch-long newborns have wrinkly, pink, translucent, hairless skin. Tiny vibrissae and claws are present, however, and the flight flap at each side is already apparent.

On the second day, some black hairs appear on the edge of the upper eyelid and just below and behind each eye, according to Donald W. And Alicia V. Linzey, who did a comprehensive study titled "Growth and Development of the Southern Flying Squirrel." The ears begin to unseal by the second day and are fully open at about the fourth day, but the young do not respond to sound until around the 19th day, according to the researchers.

At about four weeks of age, the eyes are completely open. The toes, which are fused at birth, are fully separated on the forefeet at about ten days, and on the hind feet at about two weeks, the Linzeys observed.

At six weeks of age, the new generation of flying squirrels are miniature copies of their parents. They are more than three-quarters grown, and by the time they are seven weeks old, they have increased their birth weight tenfold.

The male has no part in the rearing of his young. Apparently his role in family life ends when the female has been successfully bred. The female, on the other hand, is an attentive mother. She nurses the young several times a day, and spreads her patagium over them to provide warmth.

The tiny squirrels are born in a den lined with shredded bark and other soft material.

At six weeks of age, the new generation of flying squirrels are miniature copies of their parents.

Normally the gentlest of creatures, a flying squirrel can become a fearless defender when her progeny are threatened. She'll claw an intruder in her nest, and if a youngster falls from the den, she'll retrieve it. Time and again, foresters report the courage of a flying squirrel mother in rescuing her family from a felled tree.

Typical of the reports is an account by a Professor J. W. Stack of Lansing, Michigan: "Several forestry students were trimming trees along the river . . . when one of them sawed off a dead limb, exposing a flying-squirrel's nest. The mother squirrel left the nest, and climbed to the top of the tree. The student lifted the nest from the cavity, and carried it to the ground.

"The four young were not more than one-third grown, and their eyes were not open, and their bodies only scantily haired. The class of eight students gathered around to observe them. While they were so engaged, the mother sailed to the foot of a tree on the opposite side of the river, and immediately climbed to the top, where she made a diligent search for a suitable hole along the four large branches and trunk.

"The students were surprised next to see the parent sail to their side of the river, light at their feet, and immediately climb the pants leg of the student holding the nest and young. She grasped one of the young ones in her mouth, climbed down to the ground, went to the top of a nearby tree, volplaned (86 feet) across the river, and carried the young one to one of the holes she had previously investigated. This trip was made four times, and each time the mother lighted near the class in approximately the same spot, and also used the same starting point on the trees each time. Only once did she ascend the wrong pants leg."

When it is necessary for a mother to transport a baby, she picks it up by the loose belly skin, allowing the youngster to wrap its legs around her neck like a collar.

AERIAL DEBUT

At perhaps five or six weeks of age, the young squirrels venture out of the nest and make their first short glides, urged on by their mother. They need no instructions; they know exactly what to do and soon are masters of the skill, although some may fall a time or two in the early stages of developing their techniques.

A typical "flight" begins high in a tree with the squirrel bobbing its head from side to side, up and down (some say they are practicing triangulation). It launches itself with all four legs spread wide, stretching the patagia to create as flat a surface as possible. The tail seems to be a stabilizer. The squirrel can control its direction by raising or lowering a leg, and is able to make split-second changes in midair to avoid branches, even predators. Like that of a hang glider, its path is downward, and, depending on air currents, can be 50 yards or more, but is usually much shorter.

Just before landing, the squirrel makes a slight upward swoop by jolting its tail and head upward to reduce the speed and soften the impact. The squirrel lands upright, hind feet hitting first. Invariably, immediately upon contact, the squirrel will scoot around to the opposite side of the tree and run up the trunk, probably a technique that has been useful in outfoxing predators. It may launch off on another glide, and another, until it reaches its destination. It can cover a lot of space in a very short period of time, gliding from the top of one tree to a lower spot on another, running again up to the top and volplaning on.

The patagia, loose flaps of skin that extend from wrists to ankles, are the flying squirrel's gliding apparatus and parachute.

Just before landing, the squirrel makes a slight upward swoop.

A VARIED DIET

On their nightly forays, the youngsters begin to sample solid foods, and by about their seventh week, they are completely weaned.

Flying squirrels are basically vegetarians, with nuts, tree buds, fruits, berries and mushrooms composing most of their diet. If there is a bird feeding station nearby, they are sure to help themselves to the offerings. At night, they get no competition from the daytime feeding-station regulars—the songbirds, red and gray squirrels and chipmunks. They might be intimidated by a raccoon or opossum, but most often will have the banquet table to themselves.

The flying squirrels that visit our feeders are particularly fond of sunflower seeds, niger seeds and peanut butter. They also enjoy nibbling on the beef suet to satisfy their carnivorous cravings.

Sometimes they get a double treat when they find grubs inside acorns. When he was assistant director of the National Zoological Park in Washington, Ernest P. Walker had two flying squirrels as pets. "White grubs, so plentiful in acorns, they find particularly delicious and often select wormy acorns merely for the sake of opening the nut to get the grub," he said.

Like other squirrels, flying squirrels usually sit on their haunches to eat, holding the food in their paws. When working on nuts, they use their teeth to cut a smooth, elliptical hole in the shell to get to the meat. Gray squirrels, on the other hand, crack the nuts in their strong jaws and teeth, leaving ragged edges on the nutshells.

Flying squirrels also feed on moths, caterpillars, beetles and

At feeding stations, flying squirrels are particularly fond of sunflower seeds, niger seeds and peanut butter.

Like other squirrels, flying squirrels usually sit on their haunches to eat, holding the food in their paws.

other insects and their larvae when available. Occasionally, these innocent-looking sparkling-eyed imps will eat birds' eggs, and sometimes birds themselves, but not enough of either to be of consequence.

Noted wildlife management pioneer H. L. Stoddard is one of a number of naturalists who overlooked, or didn't realize, this characteristic and later regretted it. "On April 6, 1914, an adult female Flying-squirrel *(Glaucomys volans)* was captured with her two young, and placed in a roomy cage in the workshop, with a section of tree trunk containing a flicker's hole as a nest," Stoddard said. "Two or three days later, a fine male yellow-bellied sapsucker was captured unhurt, and placed in the same cage, where he made himself at home on the stump.

"I was greatly surprised the next morning to find his bones on the bottom of the cage, picked clean. This strong, hardy woodpecker, in perfect health, had been killed and eaten during the few hours of darkness, by the old mother Flying-squirrel, though she had other food in abundance."

Bachman, too, was surprised at the squirrel's carnivorous appetite. "On several occasions, we found it caught in box-traps set for the Ermine, which had been baited only with meat. The bait (usually a blue-jay), was frequently wholly consumed by the little prisoner," he reported.

This penchant for meat has been the downfall of many a flying squirrel and the bane of many fur trappers, who repeatedly find flying squirrels in traps set for more valuable furbearers. Some trappers feel they can't really get down to serious trapping until the flying squirrel population has been eliminated or greatly reduced at that trapping site. Some have reported snaring hundreds of the diminutive squirrels before they get their first weasel or mink.

STOCKING THE STORES FOR WINTER

Flying squirrels born in the spring are on their own by the time their mother bears her second litter in midsummer. This later litter, however, may remain with the mother as a family unit through the winter months.

As summer wanes and the glow of autumn fills the landscape, flying squirrels are busier than usual at night. They are gathering and stashing nuts and seeds for winter food. Some nuts are stored singly in a crevice of tree bark. Some are cached in tree cavities. In the case of free-running pet squirrels, nuts might turn up almost anywhere. "My pets cache nuts and other food in many different locations," wrote Ernest Walker, "on the top of a window casing, in the folds of a shower curtain, in my pockets, in the tops of my socks, inside my collar, on my arm, or in the angle of my elbow. They search out locations and tamp the nuts into place with their teeth."

Hundreds of nuts may be stored in one night, and by the time winter is in full force, niches, nooks and nests are well stocked.

THE MORE THE MERRIER

Flying squirrels are a gregarious lot, particularly during the winter. Groups of perhaps a dozen snuggle together in nests to keep warm. It seems that the farther north the squirrels live, the more individuals gather together in winter nests. They do not hibernate, but during bitterly cold spells they are inactive, preferring to remain huddled together in a cozy bundle, sometimes for a couple of weeks, until milder nights return.

Reports abound of tree cutters in the North who are amazed at the numbers of flying squirrels that scatter from a den tree when it is felled. Some have estimated that 50 individuals fled in all directions from a single falling tree.

NIGHT PROWLERS ARE ENEMIES

Flying squirrels are preyed upon by an assortment of nocturnal predators, but their major enemies are probably the large owls. Domestic cats kill their share, as do raccoons and skunks. Seton claims that an acquaintance told of finding the body of a flying squirrel in a trout!

Perhaps its greatest threat is from extensive clearcutting. Where there are no trees, there are no flying squirrels, so they are eliminated when an area is cut. Their populations also are reduced when dead trees are removed, trees which would be likely to provide nesting sites. "Leaving scattered trees, especially den trees with woodpecker holes, enables many to survive," advises Richard M. DeGraaf, principal wildlife biologist at the U.S. Forest Service's Northeastern Forest Experiment Station at Amherst, Massachusetts.

They don't hibernate, but during bitterly cold spells, flying squirrels are inactive, remaining huddled in their dens.

With flying squirrels, it's just accepted as scientific fact that they are wonderful little animals.

CONGENIAL COMPANIONS

These days, we don't encourage people to keep wild pets. In most cases, it is illegal. But in the past, those who were fortunate enough to have flying squirrels as pets praised them for their affectionate and playful nature. Even the most scientific-minded naturalists, who tend to be pedantic and unemotional in descriptions of their subjects, can't help throwing aside scholarly nomenclature when speaking of flying squirrels. German zoologist Bernard Grzimek, for example, talks of their "adorableness."

With flying squirrels, it's just accepted as scientific fact that they are wonderful little animals. "Things have gotten out of hand when someone watches fairy diddles [flying squirrels] up close, for prolonged periods, with a straight face and then reports that they are 'rodents,' " wrote Ted Williams in *Audubon Magazine*.

"We are talking here about an animal that stamps its feet when angry; lies on its back and kicks to discourage nest snoopers; embraces during mating, with the male throwing his flight cape over the female; greets members of the same denning group by literally kissing them. How," he ponders, "does one seriously record the meeting of a group member and a stranger when the subjects are dancing and shouting and all the while attempting to sniff one another's genitalia but with scant success because each is constantly slapping the other in the face with the sole of its hind foot?"

One owner of flying squirrel pets in the 1880s, Professor F. H. King, commented that "they seemed to get far more enjoyment from playing upon my person than in any other place, running in and out of pockets, and between my coat and vest. After the frolic was over, they always esteemed it a great favour if I would allow them to crawl into my vest in front, and go to sleep there, where they felt the warmth of my body . . . indeed, they came to consider themselves abused if I turned them out. When forced to go to sleep by themselves, the attitude taken was amusing—the nose was placed upon the table or other object it happened to be upon, and then it would walk forward over it, rolling itself up until the nose almost protruded from between the hind legs; the tail was then wrapped in a horizontal coil about the feet and the result was an exquisite little ball of life in soft fur which it seemed almost sacrilegious to touch. If they escaped from the cage during the night, I was sure to be warned of the fact by their coming into the bed to roll themselves up close to my face or neck."

They appear to genuinely enjoy the company of their human masters. "One will stand on my shoulder and gently bite the rim of my ear or put his nose into my ear and sniff rapidly. This seems to be a way for many animals to express affection," wrote Ernest Walker in a 1947 *National Geographic* feature about his pet flying squirrels. He added, "Some of my human friends develop remarkable agility when the playful squirrels get inside their clothes. On bare skin the tiny needle-sharp claws tickle."

The main drawback to keeping them as pets is that they are nocturnal beings. There is no way to readjust their circadian rhythms to match those of diurnal humans. They sleep soundly in their nest boxes during the day. Then, about the time you're ready to drift off to sleep, they are ready to play. If they are caged, they need to have a large wheel for exercise, and will keep it whirring almost nonstop on many nights.

Their huge black eyes, so well equipped to see in darkness, are sensitive to the brightness of daylight. One flying squirrel owner, in order to spend more time with his elfin charges, said he was trying to become more nocturnal, and used dim blue lights when the squirrels were active to avoid hurting their eyes.

They do, however, get used to the beam of a flashlight, patio light or other outdoor house lights that are turned on regularly.

THE EVENING'S ENTERTAINMENT

If you haven't seen flying squirrels at your bird feeding station, it may not be because there aren't any around. It's probably because you just didn't know they were there, or how to go about seeing them.

If you have bird feeders and mature hardwood trees in your backyard and adjacent areas, you probably have flying squirrels.

Try to illuminate the feeders at night. A very bright light is not necessary, or even advisable, just bright enough to allow you to see what's happening. Turn the light on every night at dusk and leave it on until you go to bed. If the flying squirrels have been visiting the feeders on a routine basis, they'll soon adjust to the light and pay no attention to it.

Check the feeders about once an hour each night until the squirrels make an appearance. Once you know that they are using your feeders, you can count on their returning at about the same time every night. To better observe the squirrels and not startle them with your movements, turn off the indoor lights in the room from which you're viewing them.

GLIDERS AROUND THE WORLD

In North America, there are two major species of flying squirrels, the southern flying squirrel, *Glaucomys volans,* and the northern flying squirrel, *Glaucomys sabrinus.*

The northern flying squirrel is identical in appearance to the southern flying squirrel, but is slightly larger. On the southern flying squirrel, the belly hair is pure white down to the root; on the northern, the belly hair is also white, but is slate-colored close to the skin.

Instead of inhabiting stands of deciduous trees, the northern flying squirrel is found in coniferous and mixed woodlands from the northeastern United States throughout most of Canada and into parts of California. In the southern part of its range, it is often found in higher altitudes that offer a Canadian Life Zone—the Appalachian ridges and the Sierra, for example.

Not as much is known about the northern flying squirrel, but its life-style is probably much like that of the southern flying squirrel. Because of a somewhat different habitat, its diet is a little different. It eats seeds, nuts, berries, fruits, insects, and the occasional bird egg or bit of meat, but it also consumes a great deal of pine cone seeds, meticulously stripping each seed from a cone until only the naked stem remains.

Seton, in his *Life-Histories of Northern Animals,* relates an incident in New Brunswick which shows the diversity of some northern

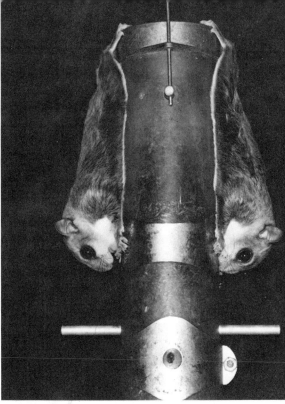

Their huge black eyes reflect red in the beam of a flashlight.

Once flying squirrels start using your feeders, they will probably return at about the same time every night.

flying squirrels' tastes: "An odd experience befell Mr. Hunter during his return from a hunting trip to the settlement last fall. One evening he left a candle burning on the table in the Forty-Nine-Mile Camp while he went out to the hovel to look after the horses. To his surprise, when he returned to the camp the candle was not only extinguished, but could nowhere be found! Mr. Hunter is not entirely free from the influence of these wild, weird legends peculiar to the backwoods of the Miramichi, especially those that related to a fabulous monster known as the 'Dungarvon Hooper.' He lit another candle, however, and went out to attend to his team. When he came back he found that the second candle had vanished as mysteriously as the first! This was a severe blow to Mr. Hunter's peace of mind but he pulled himself together and examined the camp thoroughly to see if some practical joker was not concealed about the premises. Finding no traces of anything in human form, he placed his third and last candle on the table, stood his axe within easy reach, and awaited developments. In a few minutes, a flying squirrel hopped in the door, boldly mounted the table, and knocked down the candle, thus extinguishing the flame. He started for the door with his booty when Mr. Hunter took a hand and put the rascal to flight."

More than 30 other species of flying squirrels occur in other parts of the world. One that resembles our southern flying squirrel but has a bushier tail and paler coat is the flying squirrel of northern Europe, *Pteromys volans.*

The red giant flying squirrel, *Petaurista petaurista*, of Southeast Asia is the world's largest glider. The giant flying squirrel is 3 to 4 feet long and is able to execute glides of up to 400 yards.

... K.P.H.

FLYING SQUIRREL FACTS

Description: A tiny, baby-faced squirrel with enormous shiny black eyes. Brownish above, white underneath, with a flat tail, and a loose skin flap that extends along the side of the body from the foreleg to the hind leg.

Habitat: Primarily deciduous hardwoods, especially hickory, beech, maple and oak.

Habits: Strictly nocturnal. Is very sociable among its own species, even gregarious. Gentle, and from a human standpoint, presents no problems (except for those that are unwelcome residents in attics).

Nest/Den: Usually an abandoned woodpecker hole, but frequently a natural tree cavity, and sometimes even a birdhouse.

Food: Nuts, fruits, berries, tree buds and insects. Will readily eat seeds and suet at a bird feeding station. Occasionally meat or an egg.

Voice: Has a variety of calls, from a soft twitter, to a squeak, to a churring sound.

Locomotion: Can scamper up and down tree trunks as well as any diurnal squirrel. Glides from tree to tree by spreading its patagium and volplaning. Is awkward on the ground, but may be found there occasionally.

Life Span: In the wild, they probably live less than five years, but one captive flying squirrel lived for 13 years.

NINE

AMERICAN TOAD
The Portly Caruso

"Listen," I whispered to Kit. "I hear American toads."

"Yes, I hear them, too," she responded with a sleepy voice.

It was 11:30 P.M. on a relatively warm night in April. It was the first time we had heard American toads calling this spring . . . an important event on our wildlife calendar.

Though there were only a few males singing above the chorus of spring peepers and cricket frogs, their high-pitched musical trills were unmistakable as they carried the few hundred yards from the marsh to our open bedroom window.

We enjoy the American toads we find in our backyard and garden every summer, and we love to hear them singing from the marsh. Not only does their trill mean spring to us, it is one of the most beautiful sounds in nature. One observer described the song of the American toad as resembling the opening movement of Beethoven's *Moonlight Sonata.*

Toad music has been heard by humans from the beginning of time. "In the silent world, many millions of years ago, the first sound ever produced by means of vocal cords was made by a creature resembling a frog or a toad," noted George Porter in his book *The World of the Frog and the Toad.* "The only other sounds produced by living

American toads do not give warts to people, nor are their warts contagious.

things at that time were the chirping, rasping and buzzing of insects which, then as now, are not emitted by vocal cords," Porter pointed out.

Only the male toad sings. He inflates his throat or vocal sac like a balloon, and it serves as a sounding board to intensify the volume of the song. Air enters the sac from the mouth through two slits. Toads can even sing underwater by keeping both mouth and nostrils closed, the air passing back and forth from throat sac to lungs over the vocal cords in the throat.

The purpose of the song is to attract females and, according to some researchers, to defend tiny territories in the shallow marshland. The duration of each song is 15 to 30 seconds, enviable stamina for any virtuoso. The trill rate is about 30–40 per second, reports Roger Conant in *A Field Guide to Reptiles and Amphibians of Eastern and Central North America.*

A TOAD CONVENTION

Several days later, I pulled on a pair of hip boots, gathered up some camera equipment and walked over to the marsh. It was early evening, and I had been aware of the increased volume of the toad chorus as the afternoon wore on. Not able to work any longer at my desk with the song of toads luring me, I announced to Kit, "I'm off to the toad convention in the marsh."

This was not the first place where I have been up to my hips in American toads. I had helped my father photograph them in Pennsylvania wetlands many years before.

It is typical of American toads to pay no attention to human intrusions. Unlike the frogs, which stopped singing as soon as they were aware of my presence, the toads sang on as I sloshed among them in the mucky water.

There were male toads every few feet, sitting erect, front feet straight, heads high and throat sacs inflated like great bubble-gum bubbles.

Male toads are so desperately involved with trying to find a receptive female that they will leap on the back of almost anything that moves, even other males. After I had photographed one particularly handsome guy for a few minutes, we both spotted another toad swimming past. My subject immediately leaped onto the back of the passing toad, clasping his prize with his front feet by burying them deep into the other's soft sides. An indignant chirp from the toad on the bottom apparently told my subject that he had made a terrible mistake. My subject quickly released the other male and resumed singing.

Fellow outdoor writer and friend Joe McDonald wrote about an experience he had in an eastern Pennsylvania wetland a few years ago. Joe wondered if a desperate male toad would respond to any movement, so he tested the idea by placing his hand within 18 inches

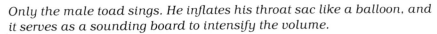

Only the male toad sings. He inflates his throat sac like a balloon, and it serves as a sounding board to intensify the volume.

of a male he had been watching. "Just as I expected, he hopped in pursuit. When the toad reached my hand he leaped aboard, scrambling with his forelimbs to attempt amplexus (embracing).

"To my surprise," Joe continued, "the back of my fist mimicked the girth of an egg-filled female and the male locked on. Even when I raised my hand the toad remained, but after a few minutes he realized something was wrong and broke his grip."

Once a male finds a receptive female, he will latch onto her so firmly that he may strangle her or injure her seriously enough to kill her. George Porter claims that he has frequently found dead females during the mating season, some of them still in the clasp of the male.

WONDERFUL FAT TOAD

Despite some problems between the sexes in the breeding season, during the balance of the year, we think the American toad, *Bufo americanus*, also known as the garden toad or "hoptoad," is simply a wonderful fat creature. It lives quietly in backyards and woodlands throughout the eastern United States and Canada. What a pleasant surprise for any gardener to lift a vegetable leaf or flower pot and find a toad quietly sitting there, looking as much at home in its environment as any creature can.

One of a dozen species found in North America, the American is a classic-looking toad, measuring a maximum of 4¼ inches nose to rump. Its rough, dry skin has one or two warts in each dark spot on its back. It also has a large, elongated gland, the parotoid, behind each eye. Large spiny warts also appear on the upper surfaces of the hind legs. Its white belly is spotted with black.

Though many American toads are plain brown, some, particularly the larger females, are brightly colored and patterned. Their ground colors vary from shades of gray to olive and brick red; their patches range from yellow and buff to orange.

The most compelling part of the American toad is its gold-flecked, jewellike eye. William Shakespeare referred to the eye of the toad in *As You Like It*. Shakespeare's toad of the English countryside, *Bufo bufo*, was considered to have magical powers by people living in the Middle Ages. It looks and behaves very much like its close American relative.

The most compelling part of the American toad is its gold-flecked, jewel-like eye.

A PASSEL OF LEGENDS

Over the ages, these much-maligned amphibians have been falsely credited with dozens of weird and scary legends. The best known and still believed is their capacity to cause warts on humans if handled.

This is associated with the belief that toads are poisonous to the touch. There is some truth to the poison theory, but they are not poisonous as far as humans are concerned.

American toads do *not* give warts to people, nor are their warts contagious. Toads do secrete a toxic fluid from their skin glands when pressured, but the toxin is harmless to humans unless it is rubbed into eyes or mouth, and then it merely causes a temporary irritation.

The poison can be very effective, however, on some predators that try to eat toads. If the predator is a dog, for example, the pooch will undoubtedly cough up the toad as soon as the poison from the toad's skin glands touches its mouth and throat. Snakes, on the other hand, are immune to the toad's defenses and often swallow them whole. I watched a garter snake do just that on our terrace a few years ago. Had the toad been full-grown, the small snake could not have handled it. Even so, the snake had a difficult time swallowing the inflated and balled-up toad. But it did manage.

Other legends claim that toads represent loathsome and hideous people such as the repulsive Jabba the Hutt in the recent Hollywood

Toads can be found in marshes while breeding in spring, and in backyards or cool woodlands in summer.

hit, *Return of the Jedi.* Still other tales claim that it sometimes rains toads and frogs, a reflection of the fact that young toads and frogs sometimes move across the landscape in great masses on rainy nights.

Finally, legend has it that unlike the frog that magically turns into a prince, toads are really witches in disguise.

SIMILARITIES OF FROGS AND TOADS

Handsome prince or ugly witch, frogs and toads do have a great deal in common. Both are amphibians which live their early (first) lives in water as tadpoles, having hatched from eggs. They have the same basic anatomy, both breed in water, both sing courtship songs and both fertilize their jelly-encased eggs externally.

The big differences between toads and frogs are their skin and where the adults live. Frogs have smooth, wet skins and long legs for leaping; toads have dry, warty skins and short legs for hopping. When not breeding, adult toads live on dry land, while most frogs spend their entire lives in wetlands.

NOT PARTICULAR WHERE IT LIVES

The dry land where adult American toads live can be anywhere from a suburban backyard to a moist woodland from sea level to mountain top. The important requirements are plenty of cover (trees, shrubs and flowers), sufficient cool, damp soil for hiding and plenty to eat.

During the heat of the day, toads like to disappear into soft and damp soil such as under a rotted log or in a burrow. If there is an extended period of very hot and dry weather, toads may aestivate in the same kind of retreat until conditions improve.

I remember being surprised to find a toad under a large slab of rock in Kit's English rock garden last summer. I had removed the rock, intending to realign it with the other rocks in the garden. An American toad was lying under the center of the slab, flat as a pancake, in a state of aestivation. As soon as I saw the dormant toad, I carefully replaced the rock in the same position.

A toad burrow is easy to recognize by its worn entrance. The occupant sits just inside the entrance and waits for dinner to pass. At exactly the right moment it lunges out, collects its food, and then backs in again for the next course. If an intruder approaches, the toad merely retreats deeper into the burrow.

To dig its burrow, which is about 3 inches in diameter and 1 foot in depth, a toad works backward, using its spurred hind feet to remove the dirt.

WILL EAT ANYTHING

An animal that will live anywhere and eat anything must be okay. Add to that the easygoing, nonaggressive personality of a toad, and you have a perfect example of a favorite species of backyard wildlife.

But there is more. Insects are very high on the list of things that compose a toad's diet. It has been estimated that during the three summer months, a single garden toad will consume up to 10,000 insects. That alone should endear this fat fellow to backyard wildlifers. Beetles, caterpillars, grasshoppers, spiders, snails, earthworms and smaller salamanders, toads and frogs are all on the toad's menu.

Toads are known to stuff themselves at times, as if it were Thanksgiving dinner. During an infestation of gypsy moths, the stomach of one toad was reported to contain 65 larvae of the pest.

With incredible speed, the toad whips out its tongue and catches its prey. As the toad swallows, it closes its eyes.

FASTER THAN THE EYE

Voracious as it is, however, a toad will never eat anything that doesn't move, claims George Porter. Nor will it stalk an insect or other moving object until the prey is almost within tongue-snapping range.

Porter writes that if the insect should stop moving, the toad will freeze, become absolutely motionless, and there will develop what appears to be a game of endurance between the two, for the toad will not move until the insect does. Then, with incredible speed, the toad whips out its tongue, which is fastened to the front of its lower jaw and is loose at the back. The insect, captured by the toad's sticky tongue, is flicked back into the toad's mouth.

As the toad swallows, it closes its eyes. This brings the eyes pressing down into the upper part of the toad's mouth, and assists in pushing the food down its throat, Porter explained.

Sometimes the toad's sticky tongue will pick up more than food. When this happens, the toad will use its front feet as hands to remove the dirt and leaves from its mouth.

There has been a great deal of research done on the tongues of toads recently. Using microcomputers, high-speed cameras and other technical equipment, two University of Michigan scientists documented that a toad's tongue flips out 2 or 3 inches, snaps into a moving snack and flips back into its mouth in less than 15/100 second, which is faster than the human eye can follow.

While primary use of the tongue is feeding, Porter points out that it also serves as a means of attack on other toads. If there is competition for food, the tongue may be snapped at a transgressor in defense or as a warning to the other to stay away.

NOT SO DUMB

Contrary to its dim-witted appearance, the toad is surprisingly intelligent for an amphibian. In fact, laboratory experiments have shown it to be the most intelligent of the amphibians (though none are very smart), states Constance Taber Colby in a *Country Journal* feature. Compared with its nearest rival, the frog, the toad reacts more promptly and learns more readily. Toads can figure out a maze far more quickly than frogs can. They discover after eight or nine trials that a glass barrier cannot be passed, whereas frogs keep bumping their noses against it. When set on a high table, toads will peer cautiously over the edge, appearing to estimate the drop, and then refuse to jump. Frogs will fling themselves off anything, Colby concluded. I've seen them do that from tables on which I was trying to photograph them.

BACK TO THE MARSH

As soon as American toads emerge from hibernation in the spring, they head for water, where they will join others of their kind to breed and lay eggs. Then they return to land. Males arrive at the breeding water a few days before females.

Not just any waterhole will do, contends R. S. Oldham in his study "Spring Movements of the American Toad." Oldham found that there is a definite homing instinct among American toads to return to the same water from which they themselves were hatched. He found that 264 individuals used the same site annually in Ontario.

STRINGS OF EGGS

A couple of weeks after my first visit to the marsh in April, I returned to check on the progress. The peak of the breeding season was apparently over. There were fewer toads, and they were singing only at night.

It didn't take me long to find what I was looking for. There were thousands of toad eggs in the exact spot where I had photographed the breeding behavior of the males earlier. They were lying in the water in bunches of gelatinous strings, haphazardly entangled among the aquatic vegetation.

STUDIO FOR BREEDERS

The strings of toad eggs reminded me of a photo session during my youth when I was helping my father record on film the breeding cycle of the American toad.

It was a rainy night in late April when we captured several pairs of breeders in a swamp close to our home in Tarentum, Pennsylvania. Hurrying back to the studio, which was set up in the kitchen, we carried the clasped pairs of toads in a wet gunny sack.

With Dad's camera focused on an aquarium three-quarters full of water, I carefully placed one pair inside. Still clasped together, they floated on the surface, apparently unaffected by their car ride in a gunny sack, the bright floodlights, camera equipment and four fascinated people watching.

Within an hour, the female began laying two strings of black eggs into the water. At the same time, the male on her back secreted invisible sperm into the water; these would swim to the eggs and enter them. It was another great experience in sex education for my sister and me.

Though we didn't count the number of eggs our pairs laid, we know that one female may lay as many as 12,000 eggs, each about 1 millimeter in diameter.

The eggs and the toads were returned to the swamp before daylight the following morning.

If water conditions are right, the tiny black specks become tadpoles in 3 to 12 days, depending on the temperature of the water.

The adults, their parental chores completed, return to dry land and head for the gardens and woodlands, where they will remain until emerging from hibernation the following spring.

METAMORPHOSIS OF A TADPOLE

Back at our Wisconsin marsh, I looked at the curled strings of eggs in the water at my feet. During this critical incubation period, many of the eggs had already turned white, either because they were not fertilized or because they had died. Only about 5 percent of those laid reach maturity, according to Porter. From the moment the eggs are laid, they are vulnerable to predation from a variety of aquatic creatures such as fish, crayfish, water beetles and dragonflies.

The female begins laying two strings of black eggs into the water while the male on her back secretes sperm.

Toad eggs lie in water in bunches of gelatinous strings, haphazardly entangled among the aquatic vegetation.

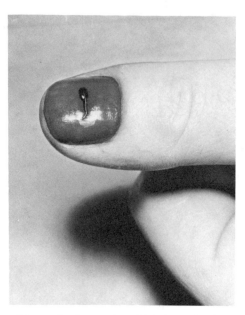

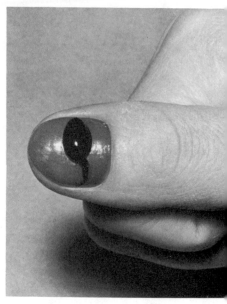

The wiggling, tiny black tadpoles, barely larger than the eggs, appear to have only a head and tail.

Tadpoles have two adhesive organs that equip them to cling to anything nearby until their organs begin to develop in a day or two.

When the eggs hatch, the gelatinous strings disintegrate. The wiggling, tiny black tadpoles, barely larger than the eggs, appear to have only a head and tail.

The world "tadpole" comes from the medieval English, *tadd* or *tade,* meaning "toad," and *poll,* meaning "head"—a toad that is all head.

Neither eyes nor mouth are evident. Tadpoles have a pair of external fishlike gills for breathing in water and two adhesive organs which equip them for their first challenge in life: to cling to anything nearby—vegetation, rocks, sticks—for a day or two until their organs begin to develop.

Tadpoles are omnivorous. Their diets include algae, particles of soft plant tissue and microscopic animals.

Tadpoles from the same parents stay together as a family. Researchers Bruce Waldman and Kraig Adler studied tadpoles and found that siblings remained together in a school or swarm, probably to decrease predation. If a predator eats one member of the school and discovers that it is distasteful, then the school is protected from further loss to that predator, concluded the scientists.

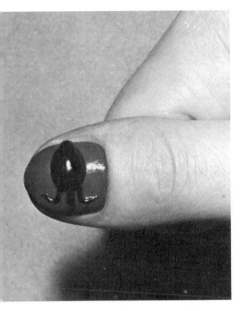

The first major change is the appearance of a pair of hind legs.

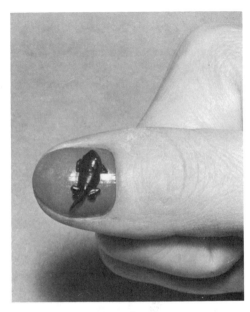

Then the front legs form.

Finally, the tail shrinks as it is absorbed and a tiny, air-breathing toadlet crawls out of the water.

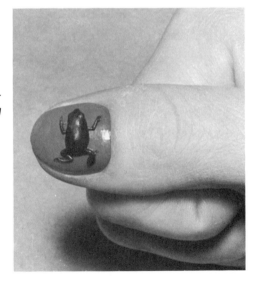

The complete metamorphosis of a tadpole from egg to toad requires 41 to 66 days. During that period, the tadpole, or "polliwog," as some call it, will grow in size and shape. The first major change is the appearance of a pair of hind legs which emerge from under the tail; then front legs form. Finally, the tail shrinks as it is absorbed, and a tiny, air-breathing toadlet crawls out of the water.

A TIME OF PENNY TOADS

Every few years when the American toads in our marsh have had a good year of reproduction, we have an invasion of what we call the "penny toads." They are swarms of tiny toadlets which have just completed their first life in water and are spreading out across the landscape in search of terrestrial habitat in which to mature.

As kids, we called them penny toads, I guess because they are about the size of a penny. Some mornings while Kit and I are walking our mile around the lake, we have to be careful where we step, for fear of trampling a penny toad.

Before another year passes, penny toads become nickel and quarter toads, and their numbers diminish dramatically from mortality.

The growing process requires the shedding of many skins. At first, young toads grow rather rapidly, shedding their skins about once a week. By the time they are nearing adulthood, they may shed only three or four times a summer.

It requires two to three years for a young toad to become sexually mature. During this time, it remains on dry land, living a solitary life, eating and growing through the warm months, hibernating through the cold.

SOME SHARP SENSES

As slow and deliberate as a toad may appear, it possesses some very keen senses. Its 24-karat gold-flecked eyes provide the animal with excellent vision.

Because the light is similar in quality and duration to that of days in the early-spring breeding season, it is not unusual to hear American toads singing in October.

Its ears, which are not really ears but circular drums located on the surface of the skin just behind the eyes, record vibrations in the air and under water. This system provides warning of approaching danger.

A toad's senses of smell and taste are not very important, though it has nostrils and reacts to some differences in scent, according to writer Colby.

Toads must have some sense of taste based on the fact that they will spit out some of what they have caught if it is not palatable, though that may be due to texture rather than taste, suggests Colby.

MANY TOAD EATERS

How a toad tastes, not what it tastes, is more important to the animal's survival. Among those predators that are immune to the toad's toxic secretions is the hognose snake, which is harmless to humans but devastating to toads. Crows, hawks, herons, owls and skunks are also immune to the toad's defenses and partake freely.

When a toad knows that it is in mortal danger, it will inflate itself to appear much larger. Then it will lower its head to form its body into a ball, creating an even more formidable mass for a predator to consume.

Skunks have been responsible for some mass killings of toads. John D. Groves, a herpetologist at the Philadelphia Zoo, reported that skunks killed and partially consumed 46 adult American toads at a breeding pool along the Patuxent River near Priest Bridge, Maryland, on April 8, 1969.

The greatest toad mortality is probably caused by automobiles. On warm rainy nights when amphibians of all kinds are on the move, millions, perhaps billions, of toads are squashed on the highways.

I will never forget the night on the island of Guam when Kit and I were driving across a rain-soaked highway en route to dinner at a friend's house. Thousands of marine toads, one of the largest toad species in the world, were crossing the highway. It was impossible to drive even a few yards without running over dozens of them. The 1-pound-plus critters literally covered the road.

Despite the toad's unique defenses, a garden toad's chances of long life are not very good. If it lives to be a toadlet, it may live an additional three to four years in the wild. In captivity, toads have lived to the ripe old age of 31.

The American toad's closest relative and near look-alike is the uncommon Fowler's toad.

THE LONG WINTER'S SLEEP

Toads that survive the summer's heat by keeping cool under logs and rocks and in burrows must retreat even deeper to escape the cold of winter.

A cold-blooded animal, the toad is well aware that winter is coming. The fewer hours of daylight and the lack of heat in the weakening sun are enough to tell the fat toad that its basking days for the year are numbered.

Because the quality and duration of light in October is similar to that of days in early spring breeding season, it is not unusual to hear American toads, and various species of frogs, occasionally singing their musical trills at that time.

Having fed to capacity on insects and other morsels during the autumn season of plenty, the toad has prepared its body for the long, cold sleep ahead. Now it is time to prepare its quarters.

At the appointed moment, which will vary from year to year depending on temperature, the toad will back into its winter den. It may be in a log or under some rocks, but most likely it will be in a mound of soft soil where it can submerge itself a foot or more into the protective insulation.

Lately, biologists have been investigating toads' and frogs' tolerances for cold. They have discovered that the animals have a kind of antifreeze called glycerol, which is also used in automobile antifreeze. William D. Schmid of the University of Minnesota has found that toads and frogs can survive having as much as 35 percent of their body water turn to ice. Antifreeze within the cells of the animal protects these vital units of organic life from being destroyed.

In spite of this protection, hibernation is no guarantee of survival, and many toads never awaken.

THE WORLD OF TOADS

There are some 200 species of toads living throughout the world. About a dozen live in North America, ranging in size from the 1-inch oak toad to the 9-inch giant or marine toad. They are all descendants of the first vertebrates which emerged from the water during the Paleozoic era to become terrestrial creatures.

The American toad's closest relative and near look-alike is the uncommon Fowler's toad, *Bufo woodhousei fowleri*, whose range overlaps the American's in the East. The two have been known to hybridize, producing offspring with characteristics of both parents. The chief difference in appearance is that the Fowler's has three of more warts per large dark spot, compared to the American's one or two.

The marine toad is one of the largest toad species in the world.

Other important toads of North America include the western toad, *Bufo boreas*, of the West Coast; the Great Plains toad, *Bufo cognatus*, found in the Great Plains, the Southwest and into Mexico; the red-spotted toad, *Bufo punctatus*, of the Southwest and northern Mexico; the southern toad, *Bufo terrestris*, and the oak toad, *Bufo quercicus*, both residents of the Southeast and Florida; and the western race of the Woodhouse's or common toad, *Bufo woodhousei woodhousei*, of the midlands and Rockies (the eastern race is Fowler's toad).

. . . G.H.H.

AMERICAN TOAD FACTS

Description: A classic garden toad, up to 4½ inches long, with warts on its back and black specks on its belly. Some, particularly the larger females, may be more brightly colored and patterned in brick red, orange and yellow. Large, elongated parotoid glands are located behind the eyes; large spiny warts appear on upper surfaces of the hind legs.

Habitat: Nearly everywhere in the East from backyard gardens to mountaintop woodlands where there is cover, moist soil and food.

Habits: Solitary, except during spring breeding season when it joins large numbers of its species in shallow bodies of water to lay and fertilize strings of eggs. Spends early life in water as tadpole. Hibernates.

Burrow/Den: Backs into burrow to hide and await prey. During hot, dry weather will aestivate under rocks and in rotted logs and loose soil. Hibernates in winter in similar locations a foot or more underground.

Food: Tadpole eats algae, particles of plants and microscopic animals; adult eats insects, snails, slugs, earthworms and spiders. In captivity, will eat mealworms. Water absorbed through skin.

Voice: Spring song of male is a musical trill, considered one of nature's most beautiful sounds. Also chirps when handled and screams when being attacked.

Locomotion: Hops.

Life Span: If it survives aquatic stage, may live three to four years in the wild. One captive toad lived 31 years.

TEN

WOODCHUCK
Underground Architect

Spending the summer in a neighbor's gatehouse a few years ago while our own house was being expanded gave us some intimate experiences with the wildlife living there at our doorstep. Warblers and vireos serenaded us from the ashes and basswoods, various butterflies flitted through the columbine and phlox, a raccoon raised her young in the sugar maple outside our bedroom door, and a family of woodchucks lived under the woodpile.

Timid at first, the chucks slowly grew accustomed to our comings and goings. They never became tame, but they weren't in quite as much of a panic to get away from us as they had been at the beginning.

The first time we were aware of them was on a warm morning in early June. We were walking across the lawn, each lost in our own thoughts, when it happened. The woodpile, about 15 feet ahead of us, seemed to rumble. Logs spilled from the top and crashed to the ground. There wasn't a sign of anyone or anything near it except us.

We continued on, and when we returned 30 minutes later, we caught one by surprise. An adult woodchuck was lying on top of the woodpile basking in the sun. Of course, as soon as it detected our approach, it disappeared under the wood in the blink of an eye.

Within the next few days, we realized that this was a female woodchuck. Her little ones started to come out to enjoy the warmth of the sun with their mother. Some days the mother and her four little clones would all sprawl out on the pile, looking a bit drowsy, we thought, like any other sunbathers.

It was a comical sight to see those small balls of brown fur in an assortment of poses. One might be on its back with its pudgy tummy sticking up; another might be sitting upright like its watchful mother; the others might be sprawled out on their bellies. No matter how relaxed or sleepy they appeared, our approach never failed to trigger a woodchuck-style Keystone Cops drill until they came to accept us. A high whistle, followed by frenzied scrambling and logs rolling in all directions, was their usual greeting for us at the beginning of our acquaintance.

FURRY BROWN BUTTERBALL

A woodchuck, or groundhog, is a furred butterball, usually 20 to 25 inches in total length, including its bushy 5½- to 6-inch tail. This marmot member of the squirrel family normally weighs from 5 to 10 pounds, but a few extremely fat autumn woodchucks have weighed 14 pounds. Males tend to be about 3 percent heavier than the females.

Their fur may be almost any shade of grizzled brown, from tawny

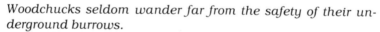

Woodchucks seldom wander far from the safety of their underground burrows.

to russet to chocolate, or any gradation of these. It covers a body that is a chunky pear shape.

The woodchuck's head is slightly flattened, with small round ears near the top. Its eyes are dark and alert, and its muzzle is white. The legs are short and powerful, and the feet have sturdy claws befitting a burrowing animal.

Woodchucks are not built for speed, except to dive into their burrows. They probably aren't capable of attaining a top running speed of more than 6 to 9 mph. For this reason, they seldom wander far from the safety of their underground burrows.

UP A TREE

Occasionally a woodchuck will climb a tree, sometimes to sun itself, sometimes as an escape when it can't get to its burrow.

Several years ago, a neighbor came rushing down our driveway. He wanted me to come with him *immediately*, because one of the local dogs was chasing a small animal in front of his house. "We think it's a beaver!" he exclaimed excitedly.

I hurried over to his front lawn to see what was going on. A beaver in our area would indeed have been something spectacular to see, but I was certain this wasn't the case.

We arrived just in time to see Sparky, the German shepherd notorious for dispatching countless neighborhood cottontails, pheasants, muskrats and chipmunks, tear around the corner of the house. She was in hot pursuit of a small brown animal that was scuttling for all it was worth across the lawn. It wasn't a beaver; it was a young woodchuck that had been unable to get to its burrow when the dog took chase.

Up a small tree it scampered, Sparky on its tail. The dog yapped and jumped at the base of the tree, but the chuck was out of its reach. (By this time, quite a crowd had gathered, everyone wanting to see the beaver.)

Then, for some reason I'll never understand, the youngster left the safety of the tree and ran down the trunk headfirst. It had barely touched the ground before Sparky sank her teeth into the nape of its neck and quickly snuffed out its life with a rough, triumphant shake.

When we finally were able to get Sparky to relinquish her prize, the others had a chance to see that the little animal had no beaver

Occasionally a woodchuck will climb a tree, sometimes to sun itself, sometimes to escape a predator.

tail. Still, some remained unconvinced, and every now and then we hear someone say, "Why, I remember when we had a beaver around here not too long ago . . . but Sparky killed it."

THESE TEETH WERE MADE FOR GNAWING

The woodchuck, *Marmota monax*, is a rodent. Rodents, quite simply, are animals that gnaw. Their upper and lower incisors grow continuously, but are constantly chiseled down to the proper level by grinding against each other as the animal gnaws.

One of the surefire ways to identify a woodchuck if you're close enough, or if you have one in hand like the one Sparky caught, is to check the teeth. A woodchuck's teeth are white. Other rodents have yellow or orange teeth.

In an estimated 1 percent of woodchucks, the teeth may be malformed or not meet properly to wear against each other, a condition called malocclusion. When this happens, the teeth continue to grow at the rate of about ¼ inch per month. The upper incisors curve downward, the lower incisors curve upward, forming long tusks. In severe cases of malocclusion, the teeth may pierce the jaw or skull, usually leading to a slow, painful death. Sometimes the animal starves because it cannot eat properly.

SENSES ARE FINELY TUNED

The senses of sight, hearing and smell are highly developed in the woodchuck. Alert to every sound, the family in the woodpile was often out of sight before we came into view. The tumbling logs were our only clue that they had been there. Their sense of smell, too, seems to be keen. Researchers and woodchuck hunters have learned that the best way to approach a woodchuck is from downwind.

With their sharp eyes, woodchucks can spot movement on a far horizon and will monitor it until they feel comfortable that it is not a threat. Or, if they perceive it to be danger, they dive into their burrows with a shrill whistle.

WHISTLES AND CLICKS

The high-pitched whistle is the sound most commonly heard from woodchucks. At the first sign of danger, a woodchuck plunges into its underground burrow while whistling this piercing alarm.

"Often this whistle is mistaken for one made by a human," according to W. J. Schoonmaker, author of *The World of the Woodchuck.* "On one occasion my wife believed that I had whistled at her from the edge of a wood lot," he recalled. "To her surprise, she found that a woodchuck had been the whistler."

Naturalists have pondered whether or not the alarm whistle is an altruistic action on the part of the woodchuck and other marmots. "Altruistic behavior may be defined as behavior that renders its performer less likely to survive while increasing the likelihood of survival

The teeth grow continuously, but are constantly chiseled down to the proper level by grinding against each other.

by another," explains David P. Barash in his report "Marmot Alarm-Calling and the Question of Altruistic Behavior."

"Intuition suggests that by attracting the attention of an observed predator, the alarm caller subjects itself to a greater risk of potential predation than if it 'selfishly' kept silent and concealed itself," he writes. For example, the alarm whistle often attracts the attention of a person who had not noticed the animal, Barash points out, and the same would seem to be true for natural predators.

However, "marmots only rarely give alarm calls while fully exposed in their meadows," Barash reports. "After seeing a predator, they invariably run first to a burrow entrance before vocalizing."

Based on what he refers to as "admittedly meager quantitative data," Barash suggests that alarm calling is not a dangerous activity for woodchucks. The animals whistling are already aware of the predator and so are somewhat protected against predation themselves.

Another common sound from woodchucks is a chattering, or rapid clicking, made by grinding their teeth when they are angry or afraid. When fighting or injured, a woodchuck will make other noises, including squeals, snarls, hisses and growls.

"Those who have kept them in captivity report that, when stroked or when requesting attention, they make a soft, purring noise which would not normally be heard in the wild," claim *Mammals of Pennsylvania* authors Doutt, Heppenstall and Guilday.

PART OF THE LANDSCAPE

Throughout most of the eastern United States and nearly all of Canada, the woodchuck is a fixture of the landscape. It occurs east

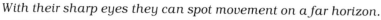
With their sharp eyes they can spot movement on a far horizon.

Woodchucks are common in rural areas, and show up with some frequency in suburban backyards and gardens.

to the Atlantic, south to Alabama and north to Alaska. In the western states, the hoary marmot and the yellow-bellied marmot fill the niche occupied in the East by the woodchuck.

It is primarily an animal of woodland borders, open fields, meadows and pastures, highway edges, brushy or rocky ravines and rocky slopes. Some find orchards and old cemeteries attractive homesites. Woodchucks are common in rural areas, and they show up with some frequency in suburban backyards and gardens, especially if there are open areas nearby.

Suburbanites and city dwellers who live near nature centers, parks or golf courses are often startled to see a rotund woodchuck sitting in their backyard next to a burrow whose entrance wasn't there the day before.

Wherever woodchucks are found, there is bound to be an abundance nearby of their favorite foods, especially clover, grass and alfalfa.

An acquaintance in a fashionable Milwaukee suburb has been challenged year after year by a woodchuck that has a particular fondness for this gentleman's tender green beans. Fencing the garden did no good, nor did repellents. The woodchuck burrows under hedges and fences to emerge smack in the middle of the bean patch. Before the man can get out his back door, the chuck is long gone.

The frustrated gardener has filled in endless chuck holes, only to have the intrepid digger pop up again the next day in about the same spot.

This has gone on for several years, and he complained to me about it recently and asked what he could do about it. I made a number of suggestions, but he shook his head at every one. Finally he said, "I guess I really don't want to do anything about it. I'd really miss that groundhog. I have to admit, I admire its determination."

ITS DAY OF GLORY

Regardless of our feelings toward woodchucks in our vegetable gardens, nearly all of us are interested in the animal when February 2 rolls around. It's Groundhog Day, the day when we supposedly find out how much more winter we'll have to endure. If the woodchuck comes out of his den and sees his shadow, we'll have six more weeks of winter. If he doesn't see his shadow, we'll have an early spring. At least, that's how the legend goes.

A charming legend, too, but that's all it is. The tradition of Groundhog Day was brought to this country by Europeans who looked to badgers or bears for weather forecasts on Candlemas Day.

Somehow woodchucks were bestowed the honor in America. The most famous of the lot is Pennsylvania's Punxsatawney Phil. Every February 2, Phil is "encouraged" to emerge from his den and make his prediction to an anxiously awaiting nation.

COURTING FIRST, EATING LATER

There may be a few woodchucks about on February 2, especially in the southern parts of their range, but many are still sleeping snugly inside their winter burrows. When they do emerge, looking for their shadow is not of primary interest, nor is finding food, even though they have not eaten since the previous fall and have lost about 40 percent of their weight over the winter. The most urgent task now is finding a mate.

The male woodchuck climbs out of his own burrow and makes a beeline to other nearby burrows. In a methodical, direct route, he goes from one burrow to another. He sniffs at the entrance, and if he detects a male inside or a mated pair, he backs off and goes on to the next one. This pattern continues until he finds a burrow that his nose tells him houses an available female. Then, with tail wagging like an eager puppy, he cautiously enters the burrow.

The female, perhaps in a woodchuck show of coyness, may charge at him and chase him out of her burrow. Undaunted, the male gingerly makes another approach, tail wagging in friendly greeting. He may get the same treatment time after time, but he is persistent, and in most cases, the female eventually relents and accepts her enthusiastic suitor.

He moves in with her for the duration of their February, March or April courtship. This is the only time of the year when two adult chucks inhabit the same burrow.

"It became evident to me that pregnant females, just before the young are born, are not sympathetic with the desires of their mates and strongly repel them and discourage their attentions," observes Schoonmaker. "The males, I noticed, when they find their mates uncooperative, go forth in search of more willing females. Their search is usually in vain because practically all females are in the same physical condition and mental states. Regardless," he continues, "the males seem not to become discouraged, and almost every day that I was in the field during the latter part of April and in May, I saw them hopefully continuing their search."

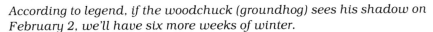

According to legend, if the woodchuck (groundhog) sees his shadow on February 2, we'll have six more weeks of winter.

A FRESH NEST

Alone again, the female does some spring cleaning in her burrow, The winter sleeping nest will become the nursery. It is a chamber off the main underground tunnel, and is usually about 15 inches in diameter and about 8 inches in height. Old nesting material is taken out and new dried grasses and weeds are brought in.

Four weeks after conception, the three to six tiny woodchucks are born. The 1-ounce babies are blind, naked, and completely helpless. They are less than 4 inches long, with a ½-inch-long tail and very short, fine whiskers. Instinctively, they seek the nourishment of their mother's rich milk and begin nursing.

Within a week, the infants double their birth weight and show a sparse covering of hair. At two weeks, the 6-inch-long youngsters weigh about 3½ ounces.

When the young woodchucks are four weeks old, their eyes open. Now fully covered with short fur, the chucks are 8 inches long and weigh about 6½ ounces.

ENTERING A NEW WORLD

At the age of six weeks, the 10-inch-long, 8-ounce woodchuck kits are active and ready to explore the brightness outside their dark den. The mother chuck first makes certain that the coast is clear. "If satisfied that all is well, she disappears into her den and immediately emerges with her babies," Schoonmaker states. "At the mouth of the den they stop and look with eyes that are soft brown and full of wonder upon a world that to them is new."

While the youngsters play, explore and taste the grasses and clover in the immediate vicinity of the burrow entrance, the mother keeps a watchful eye for predators. At the first sign of danger, she herds her family back down the burrow with an alarm call.

LESSONS OF LIFE

The woodchuck mothers we've been able to observe seemed to enjoy and be affectionate with their offspring.

A woodchuck mother must also be a good tutor, making certain that her family learns well the lessons of survival. Each time she whistles danger to them, the lesson of dashing out of sight of enemies is reinforced.

They eat what she eats, weaning themselves gradually from her milk. Woodchucks are herbivorous. Their diet consists of clover, alfalfa, dandelions, grasses, herbs, fruits and, when they can get them, cultivated vegetables and grains such as beans, peas, lettuce and corn. They'll often sit on their haunches and munch on a sprig of clover or other greenery held in their front paws.

LEAVING HOME

By midsummer, the size of the growing young woodchucks creates space problems in the home den. Now weaned, each youngster must seek its own fortune.

The dissolution of the family unit is a gradual process. Temporary burrows are found within 150 yards or less of the mother's burrow. These may be abandoned burrows, or ones that the mother has recently excavated.

For a time after the young are established in these "halfway

At the age of six weeks, woodchuck kits are active and ready to accompany their mother on excursions out of the den.

Woodchucks are herbivorous. They'll often sit on their haunches and munch on greenery held in their front paws.

houses," the mother continues to keep a watchful eye on them, sending them diving into the safety of their dens if she senses trouble. She makes the rounds daily, visiting each young chuck at its own burrow.

Her visits become less frequent as the young become more independent. Then each youngster moves again, leaving the mother's territory to establish its own. It might lay claim to an abandoned burrow or dig a new one.

MASTER DIGGER

According to an old saying, "A woodchuck eats to give him strength to dig holes, then digs holes to give him an appetite." Woodchucks are designed for digging, and they do it with amazing speed.

With windmilling front paws, the chuck throws soil with such a frenzy that it sometimes flies a yard away. When the hole deepens, the back legs work, too. While the front legs dig, the back legs push the dirt behind. When it accumulates, the chuck pushes it out the entrance. This is what forms the large mounds at the entrances to woodchuck burrows.

If a rock or tree root is encountered, the chuck digs around it if it is too large, but the strong white incisors usually gnaw through roots. Most rocks are pushed out the burrow entrance. Schoonmaker claims he once saw a 7-pound woodchuck dislodge a 15-pound rock and move it 4 feet away from the den.

In the days of horse-drawn farm equipment, farmers never welcomed the site of a woodchuck burrow. They claimed their horses or cattle would break legs stumbling in the holes . . . or even disappear down them!

Actually, woodchucks are great soil improvers, constantly bringing the subsoil to the surface. In *Mammals of the Eastern United States*, William J. Hamilton estimates that in New York State alone, woodchucks bring over 1,600,000 tons of earth to the surface each year.

The chuck works so quickly that it can easily complete a simple burrow in one day. Ernest Thompson Seton tells of a woodchuck burrow in fairly loose soil that friends dug out one day with shovels. Eventually they reached the woodchuck itself and someone was sent to bring back a canvas bag in which to carry the chuck. Meanwhile, the woodchuck decided not to wait. It began digging, filling up the hole between itself and its would-be captors, disappearing entirely from view within one minute.

In another incident, Seton says an individual was dug out again and again. Each time, the woodchuck plugged the hole behind it when exposed by the diggers. "In half a minute he could so completely close it with earth so hard packed as almost to defy discovery of the tunnel," Seton reported.

AN ENGINEERING MARVEL

Whether simple or elaborate, all woodchuck burrows have the same basic elements. No professional architect could design a better home for the chuck.

After digging inward for 3 or 4 feet, the woodchuck inclines the tunnel upward for a foot or so, then continues horizontally, perhaps to lengths of 15 to 30 feet. This technique usually prevents the tunnel from flooding.

From the main tunnel, the woodchuck creates at least two side tunnels leading to separate chambers. One is a small pit used exclusively as a toilet. The woodchuck is a very sanitary animal, depositing all of its wastes in this excrement chamber. When the pit is full, the chuck usually seals it off and digs another. Sometimes the dried excrement is removed and buried outside the burrow.

The other chamber is the nest. This room is used for sleeping, hibernating and raising young.

The very conspicuous mound at the front entrance is an ideal observation platform for the woodchuck. From there it can usually

The conspicuous mound at the front entrance is an ideal observation platform.

get a clear view in all directions. It's also a favorite place to bask in the sun on lazy summer days.

There is typically at least one other entrance, sometimes more. This is an inconspicuous hole, often in vegetation. It is dug from the inside, and the dirt is removed through the front entrance, leaving no telltale mound at the "back door." This is often an escape hole for a woodchuck, because it is very difficult to spot, even at close range.

It's not unusual for a woodchuck to drop from sight when we approach. If we know where its other entrance is, we often see the chuck peeking out at us from it a few seconds later.

MAN IS GREATEST ENEMY

Because the burrow is dug in the midst of abundant food, there is no need for the chuck to wander far from the safety of its den entrance to feed. Its home range is usually no more than a few hundred feet.

Still, some woodchucks get caught off guard and are occasionally taken by foxes. Sometimes a large hawk snatches a juvenile.

Men and dogs, however, are the woodchuck's greatest threats. The woodchuck is a popular game animal, and in many areas is considered a varmint, legally hunted at any time. Backyard gardeners and rural farmers sometimes wage war against the benign groundhog because of the location of its burrows or because of its fondness for tender green vegetables.

On the average, a woodchuck lives for only four to five years in the wild, although their potential life span may be as much as ten years.

STOKING UP FOR WINTER

With the young chucks established in their own burrows, the emphasis on food increases. During the remainder of the summer and early autumn, both young and adult woodchucks eat as much as they can hold. One chuck was found to have eaten one-third its weight in a single day. Imagine eating one-third of your weight in salad greens in one day!

This obsession with eating is a common characteristic of winter hibernators such as the woodchuck. They gorge to put on fat to sustain themselves through the months of inactivity.

By September, woodchucks are roly-poly butterballs. The adults may weigh from 12 to 14 pounds. When they scramble along on all fours, their jelly bellies actually drag on the ground.

Researcher David E. Davis determined that the weight obtained by a woodchuck at the time it begins hibernation is critical to ensure survival through the next spring. He indicates that a minimum pre-hibernation weight for a young male is about 7 pounds and for a young female about 6 pounds.

FAST FOLLOWS FEAST

Ten days to two weeks before going underground for the winter, the woodchuck completely stops eating. It is lethargic now, preferring to doze in the autumn sun.

By September, woodchucks are roly-poly butterballs, with bellies dragging on the ground.

As early as September or as late as November, the woodchuck finally waddles down its burrow, crawls into its nest and seals itself in by blocking off the entrance to its bedroom with dirt.

It curls into a ball and gradually falls asleep. As it drops into a deeper and deeper sleep, its temperature and respiration rate drop. When awake, the woodchuck normally breathes about 2,100 times an hour. In hibernation, it breathes as little as ten times an hour. From the normal 80 beats per minute when active, the woodchuck's heart slows to only four or five beats per minute. The pulse is so weak it's hard to detect. The woodchuck's internal temperature, normally 100 degrees F., gradually drops to between 37 and 57 degrees. In this comalike state, the animal seems to be more dead than alive, and will not rouse even if it is touched.

Yet, every now and then, the woodchuck will awaken briefly and then drift back to the limbo of hibernation. Many researchers now believe that hibernators must awaken at intervals of a few days to a couple of weeks to work off accumulated toxins. After a couple of deep breaths to clear their systems, they go back to sleep. Others claim that monitoring of hibernators' blood chemistry shows that there is no great buildup of these toxins.

Hibernators also wake up if their body temperatures drop too low. They have their own built-in thermostats that consistently seem to keep their body temperatures a few degrees above the ambient temperature of the sleeping chamber. If it is too cold, the animal awakens, preventing it from freezing to death.

TWO KINDS OF FAT

Only animals that can store enough food reserves in the form of accumulated body fats are able to hibernate. In addition to the usual

The woodchuck, a true hibernator, spends the winter in a comalike state in its underground den.

In spring, the woodchuck awakens and climbs to the burrow entrance to breathe the fresh spring air.

white fat which insulates and helps warm the body, hibernating animals have another kind of fat known as brown fat.

In 1551, a Swiss naturalist discovered dark-brown tissues in the upper back, neck and chest of a marmot he was dissecting. Since then, brown fat has been found in other animals, especially hibernators, and for a while was called the hibernating gland. However, it was also found in nonhibernators, and was found to be a conspicuous feature of newborn animals, including humans. Cells of brown fat produce body heat 20 times faster than white fat. The brown fat is the "electric blanket" that automatically warms the hibernating woodchuck when the surrounding temperature drops.

EARLY ALARM CLOCK

From late January to the end of February or early March, depending on latitude, woodchucks come out of hibernation. David E. Davis found that the date of the emergence of the first woodchuck of the year is about one day later for each 18 kilometers farther north.

The awakening process throws a great strain on the woodchuck's entire body. The sequence of the previous fall is reversed. Little by little, usually over a period of several hours, heartbeat and respiration speed up.

The woodchuck's body warms gradually, from the front to the rear. As the heart, lung and brain functions accelerate, increased blood flow reaches the rest of the body, hindquarters last. The chuck shivers off the final chill, drowsily digs itself out of its sleeping chamber and climbs to the burrow entrance to fill its lungs with the fresh spring air.

WESTERN COUSINS

In the West, the yellowbelly marmot, *Marmota flaviventris*, and the hoary marmot, *Marmota caligata*, are the woodchuck's counterparts.

The yellowbelly marmot, also known as the rockchuck, is the common chuck of the West. It has an overall yellow-brown coat, except for some white between the eyes. It lives among the rocks of valleys and slopes up to elevations of 12,000 feet.

The hoary marmot lives in alpine meadows and on slopes in the high mountains of the Northwest from Idaho to Alaska. It has a grayish coat, black-and-white shoulders and black feet. This is a very attractive marmot, and one that is a thrill to spot on mountain hikes.

. . . K.P.H.

WOODCHUCK FACTS

Description: A brown, chunky 5- to-10-pound animal about 2 feet long, including its bushy 6-inch tail. Its head has a slightly flattened top.

Habitat: Meadows, pastures, abandoned fields, open areas near woods, vacant lots, brushy slopes and backyards that are close to or adjacent to these features.

Habits: Diurnal. Hibernates from about October to February. Solitary except for short period during breeding season and when female is raising young. Occasionally two young woodchucks of the opposite sex will spend their first winter in the same den. Breeds when one year old.

Den/Nest: Den is in an underground tunnel system with chambers for sanitation and sleeping. The sleeping chamber is also the nest in which the young are raised. Dried grasses may be brought in as nesting material.

Food: Herbivorous. Nearly everything consumed is vegetable matter. Favorites are clover and alfalfa. Also eats grasses, herbs, cultivated fruits and vegetables, if available.

Voice: Most common vocalization is an alarm call of a shrill whistle. Also chirps, growls, snarls, squeals and hisses.

Locomotion: Usually scurries short distances. Can run, but top speed is probably less than 9 mph.

Life Span: Probably four or five years in the wild, but has the potential to live to be ten.

ELEVEN

WHITE-FOOTED MOUSE
The Whiskered Groomer

"Let's call her Molly," our daughter Jennie suggested.

"No, I think she looks more like an Agnes," I insisted.

So we compromised . . . we called her Molly.

Molly was a female white-footed mouse who would live in a hamster cage in our sunroom for a couple of months while we studied her.

She presented herself to us at a most opportune time, because we were researching white-footed mice for this book. It began one morning last summer when I noticed mouse droppings in the kitchen. "We had a guest last night," I told Kit. "It was probably a whitefoot." Though the European import, the distasteful house mouse, is common in city dwellings, the handsome native white-footed and deer mice are more apt to move into countryside buildings like ours.

That night we set a mouse-size live trap, the smallest model made by the Havahart Company of Lititz, PA, in the kitchen. The next morning, we found Molly sitting quietly in the little wire trap. We knew she was female by the nipples on her abdomen.

A trip to the local Ben Franklin store produced a hamster cage, complete with balcony and exercise wheel. After adding some cedar chips, a jar lid of cereal, a watering bottle and some Kleenex for bedding, we released Molly into her new quarters.

It didn't take her long to get acclimated to the spacious "hamsterdome." Within an hour, "Miss Priss" with the big black eyes and Mickey Mouse ears had shredded the Kleenex into a neat pile, formed it into a nest and crawled in for a snooze.

Having a white-footed mouse as a pet took me back 40 years to my youth. The son of a professional wildlife photographer, I enjoyed the keeping of many wild pets. Though they were "mine" only for the few days Dad required to photograph them, they left me with a treasure trove of memories. The whitefoots were special because they lived with us for an entire winter as my father photographed them in a variety of natural settings.

Having Molly, therefore, was like retrieving a part of my childhood. She became the focus of our lives. We even ate all of our meals in the sunroom so that we could watch her.

A nocturnal creature, our new charge did not stir again until after dark on that first day of her captivity.

A bit cautious at first, she crept out into the center of her cage, looked around, jumped up on the balcony, rippled over the top of the exercise wheel and then jumped in for a spin. Before long, the sound of the spinning wheel was one of the familiar night sounds of the house. "Racing Molly is doing her 30,000 laps again," I whispered to Kit in the middle of the night.

It all seemed to be going so well until the morning of the third day of her captivity. The moment I entered the sunroom, I knew that something was different. I could hear a strange, high-pitched squeak

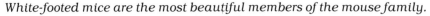

White-footed mice are the most beautiful members of the mouse family.

emanating from Molly's nest. We had not heard that sound before. Using a Chinese chopstick as a probe, I uncovered the Kleenex dome over Molly's nest and looked down at four newborn babies!

"I thought she looked a little fatter than she should have," I confessed to Kit and Jennie as they looked at Molly's infants in amazement. "This is going to be even more interesting than we had thought," I mused to myself.

"We have to give her more food and nesting material," Jennie urged, her maternal instincts showing.

Rushing around like proud relatives in a maternity ward, we added more cereal to an overflowing jar lid, and more sheets of Kleenex and newspaper bits for bedding. We were all surprised when Molly responded to this special treatment by gathering up the paper and carrying it into her nest. Jennie must have been right.

THE REAL MICKEY MOUSE

White-footed mice are the most beautiful of all members of the mouse family. Not only was the full length of Molly's 6½-inch body meticulously groomed to the tip of her hair-covered tail, she was simply one of the tidiest-looking animals I know. She weighed about 1 ounce. Males are slightly larger.

The top of Molly's head and back were chestnut-brown, with a sprinkling of dusty hairs throughout. Her underparts and feet were creamy white (thus the name). Her tail, about half her total length, was brownish-black on the upper part and white below. Her big ears were covered with a thin layer of hair; her eyes were strikingly large and very black.

Combining good looks with agility and natural grace, the white-footed mouse is one of nature's most attractive little mammals. Except for size, her appearance is not unlike that of the flying squirrel, according to some observers. (One writer reported finding the two species in the same tree cavity.)

COMMON NEARLY EVERYWHERE

The white-footed mouse, *Peromyscus leucopus*, and its near twin the deer mouse, *Peromyscus maniculatus*, are the most widespread of the more than 250 kinds of native mice in North America.

Combining good looks with agility and natural grace, the whitefoot is one of nature's most attractive little mammals.

From the Arctic Circle south, there is barely an acre that is not populated with one of the 15 subspecies and 75 races of these handsome mice.

They are also known as woodland mice and in some regions as vesper mice because of the buzzing or trilling they make by drumming their feet. This sound is a far cry from the classic "squeal" most people associate with mice. Whitefoots produce the high-pitched squeaks or squeals of terror when they are trapped or caught by a predator. While eating contentedly, they make a churring sound.

On rare occasions a whitefoot will sing a high-pitched song like a bird. The mouse's "song" was documented by a Mr. Hiskey in an early-20th-century issue of *American Naturalist:* "I was sitting . . . not far from a half-open closet door, when I was startled by a sound issuing from the closet, of such marvelous beauty that I at once asked my wife how 'Bobbie Burns' (our canary) had found his way into the closet, and what could start him to singing such a queer and sweet song in the dark. I procured a light and found it to be a mouse! He had filled an overshoe from a basket of popcorn which had been popped and placed in the closet in the morning. Whether this rare collection of food inspired him with song I know not, but I had not the heart to disturb his corn, hoping to hear from him again. Last night his song was renewed. I approached him with a subdued light and with great caution, and had the pleasure of seeing him sitting among his corn and singing his beautiful solo. I observed him without interruption for ten minutes, not over four feet from him. His song was not a chirp, but a continuous song of a musical tone, a kind of to-wit-to-wee-woo-woo-wee-woo, quite varied in pitch."

A WOODLAND SPIRIT

Because it is a nocturnal animal, most backyard wildlifers are not aware of how common white-footed mice are until one ventures inside, as Molly did. The whitefoot's natural habitat is the forests and brushlands wherever there is a good supply of food. It finds cover during the day in logs, burrows and brush piles, and quite often inside dwellings, particularly cabins, cottages, hunting camps and suburban homes.

For many years we owned a log cabin along Standing Stone Creek in Huntingdon County, Pennsylvania. A part of the adventure of being there was living with the wildlife. Upon our arrival at Hidden Valley, it was routine to remove all the mouse nests from the dresser drawers and pantry shelves. At night we fell asleep listening to the rustling of tiny white feet racing across the false ceiling above our beds. And what a surprise when a pair of beady eyes and big ears looked back from behind a soup can on the pantry shelf.

A similar experience is described in Hartley H. T. Jackson's *Mammals of Wisconsin:* "At night they almost took over our crude camp . . . and frequently scampered over our sleeping bags and sometimes over our faces. They did not bother our food supplies seriously, probably because we left exposed a liberal supply of rolled oats for them."

At the authors' Pennsylvania cabin, it was routine upon arrival to remove all the whitefoot nests from the dresser drawers and pantry shelves.

WILL EAT ALMOST ANYTHING

Rolled oats was a favorite of Molly's, as were raisins, lettuce, carrots, peanut butter, apples and a little raw hamburger. She also liked the commercial mixture of dried hamster and gerbil food.

In the wild, whitefoots eat seeds, nuts, fruits, berries, mushrooms and insects. If you ever find an acorn shell that has been cleaned out and resembles a thimble, it is most likely the work of a whitefoot.

As natural food becomes more abundant through the summer and autumn, white-footed mice accumulate great caches of food in preparation for winter. They do not hibernate, and therefore require a steady food supply to keep them going through the cold months.

With internal cheek pouches (not near the capacity of chipmunks'), whitefoots hoard the seeds of black cherry, poplar, hemlock, sycamore, maple, birch, black gum, wild grape and other trees and plants. Autumn inspections of our Pennsylvania camp always revealed hoards of white-footed food with their nesting materials.

William J. Barry found that mice hoard more food in colder climates than in warmer ones.

A whitefoot will sometimes drop its hoard down the spout of a tea kettle or into a milk bottle, with apparently no thought of how it will retrieve it, according to *Mammals of Pennsylvania*. Soap is eagerly gnawed on, hairbrushes are shorn of their bristles, tin-can la-

An acorn shell that has been cleaned out and resembles a thimble is most likely the work of a whitefoot.

bels are shredded, and mattresses are explored, all with those sharp, busy little teeth.

Backyard wildlifers who maintain bird feeding stations have to store their supplies of seed and other foods in cans with tight lids, lest they be burglarized by Molly's kin.

The damage caused by the rascals would be somewhat more tolerable if those being victimized could enjoy the comical sight of whitefoots eating in their typical posture of sitting while holding food in their front feet.

SMALL HOME RANGE

As active and busy as Molly appeared, we learned that whitefooted mice rarely travel more than a few hundred feet from where they are born. According to E. Laurence Palmer's *Fieldbook of Mammals*, of one group of 675 young deer mice studied, 70 percent of the males and 80 percent of the females settled within 500 feet of their birthplaces.

During the first two or three hours after dark each night, the animal is on the move in its territory, searching for food. We noticed that Molly ate more during that period than any other.

Thus, their ranges or territories are quite small—a single dwelling, a barn, a small patch of woodland. W. H. Burt found that the home ranges of male white-footed mice are .16 to .54 acre and those of females .06 to .37 acre.

Backyard wildlifers who maintain bird feeding stations have to store their seed supplies in cans with tight lids to keep out the whitefoots.

Mice generally live alone during the winter, but as spring approaches, males begin to seek mates.

ONE PLUS ONE EQUALS FOUR

Mice generally live alone during the winter, but as spring approaches, males begin to seek mates. In the north, the breeding season starts in February and lasts until November, according to Richard M. DeGraaf, one of the authors of *Forest Habitats for Mammals of the Northeast.* During the first part of the courtship, the females often resist the advances of males and are likely to drive them away. But soon the roles are reversed, and it is the female who winds up enticing the male into her nest, where he remains a few days to breed and then leaves or is driven away.

During the 22- to 25-day gestation period, the female prepares her nest to receive a litter averaging four young. The nest is a globular affair, usually placed 4 to 10 feet above the ground. It may be a roofed-over bird nest, a swatch of Spanish moss or inside a birdhouse, an old stump, a log, a stone wall or an underground burrow. One was reported in an abandoned hornet's nest. The female carefully lines the nest with shredded plant materials to insulate it.

At Ohio State University, researchers Harriet Glaser and Sheldon Lustick found that temperature also directly affects the amount of insulation mice place in their nests. In other words, if the weather is warm, they build thinly insulated nests, and if cold temperatures prevail, they construct warmer nests.

Molly seemed to use all the Kleenex and newspaper we provided.

VISE-GRIP HOLD

We were fascinated by Molly's four naked, blind, pink-colored babies. They looked like squirming pink grubs with whiskers.

All had attached themselves to Molly's nipples and held on with a vise grip. We discovered that this is a protective measure to allow the babies to hold on should a predator cause the mother to flee from the nest. Their grip is so tight on the mother's nipples that the youngsters are literally dragged right out of the nest and remain attached to the female in her flight from danger.

We took the liberty of looking at Molly's kids many times each day. After all, this was a rare opportunity to learn first hand about the nest life of the white-footed mouse.

The growth of the young was amazingly rapid. We couldn't believe their development from one day to the next. By the end of the second day, we could see hair growth on the babies' backs.

We found that Molly was a typical devoted whitefoot mother. During the first week or so, she remained in the nest with her litter, leaving only for short periods at night to eat and drink. Otherwise she was nursing them, licking them, and keeping them warm. Though she had no need to worry about predators in the hamster-dome, instinct made her ever watchful and protective. Even our pres-

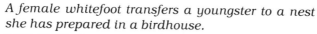

A female whitefoot transfers a youngster to a nest she has prepared in a birdhouse.

ence constituted a threat, though as time passed, she showed less concern about us. The bigger the youngsters grew, the more noise they made.

By the end of their first week, the young were nearly fully furred and several times larger than at birth. That mouse milk is powerful stuff!

TWO-WEEK WONDERS

The kids opened their eyes at the end of the second week, and they were venturing out of the nest a few days after that. Unsteady at first, they moved around on wobbly legs, but managed to get back to the nest after short outings. By then they were nearly half the size of their mother and completely covered with silky gray fur. They retain their distinctive gray coats until they are sexually mature at the ripe old age of just six to seven weeks.

With each passing day, we noticed a greater amount of activity in the hamsterdome. At night, Molly was spending much more time out of the nest, and the mouselings were exploring its far corners and finally the exercise wheel.

We watched for signs that the youngsters were being weaned. Sometime after the third week passed, we saw them eating hard food. About that time, Molly became less and less attentive. In fact, she appeared to be trying to get away from them a great deal of the time. That was when we decided to return to the Ben Franklin store for a second hamsterdome.

Transferring them from one cage to another became an Olympic effort. All, including Molly, escaped into the sunroom. Fortunately, the sunroom had no holes or cracks to the outside or to other parts of the house. It boiled down to a mighty race around and around the sunroom, with me on my knees chasing gray missiles flying in all directions. We finally got them settled into their new digs.

NOW THERE WERE THREE AND TWO

With three youngsters in one cage and Molly and a young male in the other, life settled down to a routine for the next month or so. At night we could hear the exercise wheels whirring at top speed. There seemed to be a great deal of churring and squeaking all night long,

At the end of the second week, whitefoot youngsters' eyes open, and they venture out of the nest a few days later.

The growth of the young is amazingly rapid.

and in the morning, not a trace of a mouse. By the time we awakened, they were sound asleep. The only evidence of the all-night party was disheveled cages and the floor under their cages strewn with shavings, food, Kleenex, newspaper and droppings.

CLEAN BODIES, DIRTY HOMES

It suddenly became apparent that whitefoots are more concerned about their personal cleanliness than about the sanitation of their homes. It is a fact that the white-footed mouse is among nature's most meticulous groomers. It was not unusual to watch Molly spend 15 to 20 minutes washing herself from whiskers to foot. As Will Barker states in his book *Familiar Animals of America,* "A white-footed mouse washes its face and long, sensitive whiskers much like a cat washes. After this part of the body is gone over with painstaking care, the animal licks and smooths its fur. . . ."

It is ironic, therefore, that an animal so neat about its body is so sloppy about its home. Barker continues, "It does not bother to leave its nest to go to the toilet. In a short time, the nest . . . becomes so foul that the inhabitant has to move to another."

This was certainly the case with Molly and her kids. We had to change the cedar shavings and floor paper in the cages several times a week. We had to sweep the sunroom floor every morning.

TIME TO GO FREE

With this increasing nuisance factor in mind, plus the fact that autumn was upon us, we decided to release the whole family. We had learned a great deal about the white-footed mouse, its life-style and its young. There didn't seem to be any logical reason for keeping them any longer.

So, one warm, clear evening in late August, I placed both cages on the front terrace and opened the doors. The next morning, the mice were gone.

If Molly and her kids lived normal white-footed mouse life spans, only one of them survived more than a year, and it will probably never see its third birthday.

All flesh-eating creatures hunt mice. Millions—perhaps billions—are eaten every day in North America.

WHITEFOOTS ON MANY MENUS

The white-footed mouse is a "buffer" species. That means that the animal is food for a great many other creatures. Therefore, the chances that Molly and her offspring would be around a year later were slim.

All flesh-eating creatures hunt mice. That includes predatory birds, such as hawks, owls, crows, even blue jays and shrikes. Weasels, foxes, coyotes, bobcats, shrews and snakes all consider mice a staple. Literally millions, perhaps billions, of mice are eaten in North America every day. It's a good thing, too, because the reproductive capability of mice is phenomenally high. Considering that whitefoots reach sexual maturity in six to seven weeks, and that each female can produce four or more litters a year, it wouldn't take long to be knee-deep in mice were it not for natural controls such as predators and parasites.

Mice, like other rodents, experience cycles in population numbers. Every few years, the population will grow to the point of exceeding the carrying capacity of the habitat in which it lives. At this point, natural controls, such as starvation, abundance of predators, rampant disease and/or parasites, will start the population into a downward spiral. Thus, mouse numbers (or the numbers of any form of life) never get totally out of control in nature.

Even biologists have a difficult time seeing the difference between the white-footed and the deer mouse where their ranges overlap.

THE OTHER MICKEYS

The white-footed mouse is not alone in the world of mousedom. The closest relative, the deer mouse, *Peromyscus maniculatus*, is so close in appearance and habits that it has been treated as the same species in this chapter. Even the biologists have a difficult time seeing the difference between the white-footed and the deer mouse where their ranges overlap, which is in most of the East, South and Midwest. The wider-ranging deer mouse has a bicolored and somewhat shorter tail.

There are several other members of the *Peromyscus* genus. They include the cactus, Merriam, California, canyon, brush, piñon, rock, white-ankled and pygmy mice of the Southwest and the oldfield, cotton, Florida and golden mice of the Southeast.

There are also grasshopper mice of the genus *Onychomys*, which inhabit the prairies and southwestern deserts. Their short, white-tipped tails separate them from *Peromyscus*.

The harvest mice, *Reithrodontomys,* are smaller, resembling the house mouse. They inhabit the Southeast and Southwest.

Like so many wildlife pest species, the house mouse, *Mus musculus,* is not a native. Originally from Asia, it reached North America aboard ship from Europe in the 16th century. Today it thrives in urban buildings throughout most of the continent. Its dull-gray fur and long scaly tail distinguish it from the more attractive North American natives.

. . . G.H.H.

WHITE-FOOTED MOUSE FACTS

Description: A handsome 5- to 9-inch mouse (tail is half its length) with fawn to chestnut-brown back and white undersides and feet. Large ears and large, beady black eyes. Weight averages less than 1 ounce.

Habitat: Brushy woodlands, streamside thickets, suburban and rural backyards, farms and farm buildings.

Habits: Nocturnal. Solitary, unless breeding or raising young. Home range about ¼ acre. Hoards food for winter.

Den/Nest: Globular, often in old bird nest, birdhouse, stump, log or stone wall. Made of grass and sticks and lined with shredded plant materials.

Food: Seeds, nuts, fruits, green plants, insects and small amounts of meat (carrion).

Voice: Churring, squeaking, squealing, and high-pitched singing. Will also make a buzzing sound by drumming its feet.

Locomotion: Gallops in a zigzag manner, though more often a walk or trot. It also stalks and pounces on prey like a cat.

Life Span: Less than one year in the wild, though capable of living a year or two longer. In captivity, eight years.

Box turtle habitat is woodlands, edges, thickets, marshes, bottomlands and backyards.

TWELVE

BOX TURTLE
A Four-Legged Tank

"My turtle will beat your turtle," I bragged to my boyhood buddy Dwight Chapman.

"No he won't," Dwight contested.

"Yes he will," I insisted. "I've seen yours move, and he really doesn't move very fast," I fumed.

Thus, the mighty turtle race of 1944 was on . . . and it was a fight to the finish.

Well, there really never was a finish, but my turtle, "Boxy," at least moved a couple of feet, while Dwight's, "Patton," didn't even come out of its shell.

In truth, neither box turtle was much interested in going anywhere. They are characteristically slow-moving animals that rarely travel more than a few hundred yards in a lifetime—which can be 60 to 80 years—from where they were hatched.

Box turtles typically have a home range no larger than a suburban backyard, a wood lot or a thicket. Richard M. DeGraaf and Deborah D. Rudis stated in *Amphibians and Reptiles of New England* that the home range of the eastern box turtle is 150 to 750 feet and that they retain the same ranges for years. According to researcher

J. A. Allen, one box turtle was found within ¼ mile from where it was released 60 years earlier.

I can believe this, because old Boxy lived for a good many years in and around my grandmother's backyard in Natrona Heights, Pennsylvania. It was her house that I was visiting when Dwight and I played together. Boxy was usually down in the thicket at the far corner of her property.

A UNIQUE REPTILE

The most often-kept of all wild turtle pets, the eastern box turtle, *Terrapene carolina*, is a dry-land turtle with a colorful, high, dome-shaped shell measuring 4 to 8 inches across. Both upper (carapace) and lower (plastron) portions of its shell are sectioned into 30-some plates, which are spotted and streaked with patterns of yellow, orange or olive on black or brown.

Males can be differentiated from females by their concave plastrons and sometimes red eyes; females have flat or convex plastrons and usually brown eyes.

Like other reptiles, the box turtle has scales covering portions of its skin. Its legs are covered with scales, which overlap and are

The eastern box turtle is a dry-land turtle with a colorful, high, dome-shaped shell.

The shell is sectioned into 30-some plates, which are spotted and streaked.

tougher than on other parts of the skin. Its toenails, used for digging and shredding food, are hard extensions of its skin.

Box turtles have four toenails on each front foot and either three or four toes on each hind foot, depending on the subspecies.

IT'S ALL IN THE SHELL

The shell is the turtle's most unusual feature. Obviously it gives the animal a great deal of protection against a host of would-be predators. A box turtle's shell can support a weight 200 times its own.

But the turtle's shell is a feature that is not very well understood. Even some people who have had turtles as pets think of the shell as an outside armor, protecting the animal inside. They do not realize that the shell *is* the animal inside. "The turtle is one with its shell," states Richard E. Nicholls in his *Book of Turtles.* The shell is the outer layer of the turtle; its bone structure is fused to the shell and the internal organs are beneath the shell. A turtle cannot leave or shed its shell any more than we can step outside our skins, Nicholls explains.

Turtles first developed a shell—back in the days of dinosaurs—simply to survive. Obviously it was a good move for the turtles, because while dinosaurs and many other forms of life have come and gone over these 250 million years, turtles have continued to survive.

That is not to say that turtle shells have remained the same all those eons. To the contrary, they have adapted to the ever-changing environment of the earth. That, perhaps more than for any other reason, is why turtles are the oldest living backboned land animals on earth today. They are still relatively common, with more than 200 species roving the oceans, trudging across hot desert sands and stalking the woodlands and wetlands of every continent except Antarctica. Of the 50 species in Canada and the United States, 80 percent are freshwater and land species. The remaining 20 percent are endangered sea turtles, which come ashore only to lay eggs in the sands of remote beaches.

When threatened, the box turtle draws in its head, feet and tail and shuts its shell . . . tight as a box.

A DOOR-SLAMMING SHELL

The hallmark of the box turtle is a broad hinge dividing the plastron into movable front and back sections. These doors allow the turtle to close its shell—as tight as a box. When threatened, the box turtle draws in its head, feet and tail with a hissing noise as air is released from within, and shuts its shell. No other turtle has this unique feature. In fact, it can close its shell so tightly that a knife blade cannot be slipped between the sections (not to mention the teeth or claws of a predator) to separate them.

I remember how Boxy would slam shut tighter than a drum the first time or two that I picked him up after not handling him for some time.

BOXES ARE WIDESPREAD

Strictly North American, box turtles range widely over the eastern and central United States, as well as in the Southwest to northern Mexico.

The hallmark of a box turtle is a broad hinge dividing the plastron into movable front and back sections.

Each turtle seems to be an individualist, displaying unique personality traits. John T. Nichols spent over 20 years marking, measuring and observing box turtles on Long Island. In a *Pennsylvania Game News* article, writer Chuck Fergus told how Nichols paid special attention to the demeanors of the box turtles he met. Some flailed their legs incautiously when he lifted them by the shell; others shut their plastrons tightly; a few tried to bite. Over the years, Nichols found that their personalities changed very little. Out of one group of 24 box turtles, Nichols claims that only two failed to behave in the same manner during successive recaptures. One male, Nichols admitted, changed from "retiring" to "bold." A female went from "retiring" to "bold and somewhat restless."

A LONELY LIFE FULL OF TURNS

Diurnal and solitary, except during mating, box turtles are not social creatures, though their home ranges do overlap.

Their daily routines involve a slow shuffle across their domains in a rather haphazard fashion, turning, doubling back and crisscrossing their own paths, cited Lucille F. Stickel in her study "Population and Home Range Relationships of the Box Turtle." Stickel trailed box turtles around 29 acres at the Patuxent Wildlife Research Station in Laurel, Maryland, from 1944 to 1947. Using white thread played out from spools taped to the backs of box turtles, Stickel determined that the average home range was 330 feet in diameter for males and 370 feet in diameter for females.

The box turtle's daily routine involves a slow shuffle across its domain.

FITTING INTO FORM

Stickel also found that box turtles spend the night in forms which they dig themselves by pushing soil and litter aside with their front feet and then sliding into the recess. Partly covered, but with dome sticking out of the litter, they relax and fall asleep. If the weather is particularly dry, they may spend several weeks in their forms or under logs or decaying vegetation, aestivating. If the weather is particularly hot, they may soak themselves for hours or days in mud or shallow water.

LIVE CLOSE TO WATER

Though terrestrial, these turtles prefer to live close to wetlands. That means woodlands, field edges, thickets, marshes, bogs, stream banks, forest bottom lands or backyards which have some source of water.

When conditions are right, box turtles spend most of their daylight hours searching for food. Emerging from their forms as soon as they are warmed by the sun, they shuffle off on their toes, aimlessly plodding along until they find something to eat. With each waddling step, they canvass the few feet ahead of them for such delicious morsels as earthworms, slugs, snails, insects and their larvae (particularly grasshoppers, moths and beetles), crayfish, small frogs, toads, snakes and carrion. Vegetable matter includes leaves, grass, buds, berries and fruits. They seem to love mushrooms. Researchers De-Graaf and Rudis state that young box turtles are chiefly carnivorous while the oldsters are more herbivorous.

Boxy, Patton and other box turtle pets I have known all consumed a mixture of meat and vegies. They readily munched on the leaf lettuce, fruits and earthworms we gave them. A little chunk of hamburger was also welcomed at times. We always kept their containers of water filled with clean water, because they seemed to spend a great deal of time in them.

Turtles have no teeth, but they do have hooked beaks and sharp ridges around the edges of their mouths for biting, cutting and tearing.

HIGHLY DEVELOPED SENSES

Contrary to what might be assumed of a slow-moving, primitive-looking creature, turtles have highly developed senses. Their acute senses of smell and sight help them find food and mates. They also

Box turtles spend most of their daylight hours searching for food.

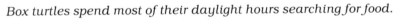

have a keen sense of touch, which they use in blindly digging nests by feeling with their hind feet, Nicholls tell us. Only the keenness of their sense of hearing is questionable. Instead, they apparently have an unusual ability to sense the vibrations of approaching danger.

The only vocalization they make is in the form of hissing, which is not actually vocalization, but the sound of air being released from within the shell while it is closing.

SEX AND THE SINGLE TURTLE

Finding a box turtle in the backyard, in a fallow field, or along a lakeshore or creekside is a fairly common event for people who spend time in the outdoors. To find a pair of box turtles breeding is rare.

It happened only once to me. I was exploring the creek bottom on my grandfather's farm in western Pennsylvania as a teenager many years ago. It was one of those gorgeous spring days in April. While walking along looking for morels for supper, I spotted the two in the grass, quietly fused together in copulation. After taking a close look, I left them alone. Had I been a little sooner, I might have witnessed the three phases of courtship.

Soon after emerging from hibernation, the male begins searching for a female. The meeting of a male and a female is not much more than mere chance.

Approaching the female to within a few inches, the male will assume a prominent pose, with his head erect and legs straight, as if announcing his intentions. Then the male will circle the female, still on straight legs, gently nipping and biting her legs as he passes. He parades around her for a while and finally, as he becomes bolder, nudges her shell with his. If she continues to show interest, he mounts her, hooking his hind toes into the back edge of her plastron. The male's concave plastron accommodates the female's carapace. Once they are locked together, he slides backward into an almost upright position to achieve copulation. They may remain locked in this position for several hours until insemination has been completed.

Eventually, the female will suddenly move away, leaving her mate in a rather precarious position, which often results in his landing upside down . . . a predicament that could result in his demise if he lands in terrain where he cannot right himself.

THE NEST IS NEXT

At this point the male has fulfilled his role and will have no further responsibility toward his offspring.

The female, however, still must perform some important tasks before her duties are complete. In a matter of weeks (there are records of turtles having laid fertile eggs several years after breeding, suggesting that sperm can be stored), the female will find a suitable soil embankment, sandbank, sawdust pile, railroad bed or garden mound into which she will dig a flask-shaped nest, usually under the cover of darkness.

The nest can only be as deep as the full length of her hind legs. The webs between her toes help her to scoop out the soil. She alternates digging with one foot, then the other. The excavation project usually takes several hours. If she is interrupted, she may quit and try again the following night in a different location.

MEMORABLE NIGHTS AT THE NESTS

I have watched several species of turtles dig their nests and lay their eggs. It was always a great nighttime adventure, either as a boy

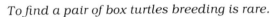

To find a pair of box turtles breeding is rare.

When the female finishes digging her flask-shaped nest,
she deposits her eggs, as shown in this cross section.

helping my father photograph the events or as an adult taking both photographs and notes for future writings.

One of the most memorable was the night that I helped my dad photograph a box turtle as she dug her nest and laid her eggs in our neighbor's terrace. We got a call from our excited friend about 10:00 one warm evening in mid-May. His dog had found the female box as she was digging her nest. Apparently undisturbed by the intrusion, she was hard at it when we arrived with camera and flashlights. For the next couple of hours, we watched as she dug the nest while half the neighborhood watched.

THE FINAL ACT

When the female box turtle literally feels with her hind feet that the nest is just right, she begins to lay her four to six flexible, pink, oblong eggs, which are smaller than Ping-Pong balls. Egg laying can be accomplished in a matter of minutes.

Following the deposit of the last leathery egg, she begins to scrape the soil back into the hole in much the same rhythm she employed to remove it, with one hind leg at a time. But the burying process is done more quickly, as the eggs are vulnerable to a long list of predators. Skunks, raccoons, opossums, rats and dogs are all interested in eating turtle eggs. Apparently, the eggs are not difficult for some predators to detect even days or weeks after the female turtle has covered them. I have returned to any number of turtle nests to find that a predator has been there ahead of me. All that is left is curled pieces of leathery shells.

When the hole has been filled in, the female then tamps the dirt with her feet and shell and tries to camouflage the disturbed soil by passing over it several times. Then she leaves, never to return . . . never to see her offspring.

IN THE HEAT OF THE SUN

If all goes well, the heat of the sun will incubate the buried eggs and the embryo box turtles will develop inside their leathery encasements. In about three months (87 to 89 days), the baby turtles will hatch. Aided by an egg tooth on their beaks, they break out of their shells, dig to the surface and disperse over the landscape.

Years ago, after the female box turtle on our neighbor's terrace covered her nest and waddled away, we removed the eggs from the nest and reburied them in a large coffee can filled with soil. The can was then placed in a sunny window at home. We were careful to keep the soil moist, but not wet.

When the time was about right for hatching, I was assigned the job of monitoring the eggs daily until they began to hatch.

The sight of baby turtles lunging out of their shells was thrilling. Soaking wet, they were immediately on the move, scurrying for any kind of cover available on the countertop where Dad was photographing them.

Fledgling turtles are at greatest risk the moment they surface. In addition to all the four-legged predators mentioned above, winged enemies, such as crows and gulls, are also on the alert to eat tender young turtles. Baby box turtles have not yet developed their hinged plastrons and cannot retract their tender limbs into the shell and close the door. Even if they could it would not help much against sharp teeth and bills, because their shells are very soft.

Unlike most wildlife, young turtles do not grow rapidly. They average only about ½ to ¾ inch a year for five to six years. Five to ten years are required for box turtles to become sexually mature, according to researcher S. A. Minton, Jr., in *Amphibians and Reptiles of Indiana*.

Box turtle hatchlings are at greatest risk the moment they surface.

Baby box turtles have not yet developed their hinged plastrons and cannot retract into their shells and close the door.

THE SHELL IS THE MIGHTY PROTECTOR

Older box turtles have less to fear from predators. Their shells do protect them against most, though I have seen big dogs try to defy this fact by "worrying" a box turtle for hours, trying to get it to come out of its shell.

Nature writer and friend Leonard Lee Rue III described a confrontation he witnessed between a box turtle and a raccoon: "I saw a raccoon discover a box turtle that had just lumbered out on a beach from high weeds on the river bank. Sensing a free meal, the 'coon hurried over to investigate. Grabbing the turtle, it bit at the head, which was promptly withdrawn. Not to be thwarted, the raccoon felt for one of the turtle's legs and tried to grab it, but the turtle pulled it in. The raccoon tried for the legs on the other side, and these, too, were withdrawn. At this point the turtle became tired of being poked at, so it slammed its shell shut. The raccoon, not about to give up so easily, turned the turtle over and over like a plaything. At last convinced that it had been outmaneuvered, the raccoon dropped the turtle and went on about its business."

VULNERABLE TO CARS

As effective as the box turtle's shell is against predators, it is no match for cars. Millions of turtles are killed on our highways each year as they cross them in search of food, mates and nesting sites.

How often I have seen a black spot on the highway well ahead of me as I have been driving over interstates in the East. Guessing what it was, I have usually managed to miss hitting the high-domed, yellow-marked carapace of the eastern box turtle as the blurred round form disappeared under the front bumper. But I have always looked into the rear-view mirror to make sure.

As the bump on the concrete quickly disappears into oblivion, I am reminded of that great chapter in John Steinbeck's classic *Grapes of Wrath* in which a turtle crosses a highway and survives. I always wonder if the turtles I have just passed will be as lucky.

AGAINST THE COLD

Because turtles are cold-blooded, they have no way to keep warm during winter except to hibernate.

With the shorter periods of daylight and the lack of warmth from the sun, box turtles living north of the freeze line must prepare for the long inactive period ahead. Having indulged in the bountiful foods of late summer, box turtles store fat to fuel their dormant bodies over the winter.

On some mysterious signal in September or October, box turtles will dig deep with their front feet into the woodland floor, or a rotted log, a stream bank or an earthen mound in a garden where they will hibernate. After digging down about 2 feet and backfilling to cover themselves, box turtles will partially close their shells, relax their feet and heads and then close their eyes. The long sleep begins.

Occasionally, warm weather in January or February will give the hibernating box turtles a false signal to emerge. Some will be caught out in the cold and perish.

Hatchlings which were conceived late in the summer may not emerge in fall. Instead, they will hibernate in the nest over winter.

After digging down 2 feet and backfilling to cover themselves, box turtles begin hibernation, as shown in this cutaway photograph.

Rather than confining a pet turtle to a cage or enclosure, why not seed your yard with a couple of free-ranging box turtles?

BOX TURTLES IN THE BACKYARD

Because box turtles are very adaptable creatures, any backyard with sufficient cover, food and water to attract birds will support a box turtle or two.

Among all the backyard wildlife in this book, none is better suited as a pet. Their adaptability, quiet nature, and longevity all add up to a nice pet which is easy to care for.

But rather than confining a pet turtle to a cage or enclosure, why not simply seed your yard with a couple of free-ranging box turtles? Assuming that you have found the turtle in the general area of your home, conditions in your yard should be suitable, as long as there is cover, food, and water.

Pennsylvania Game News writer Chuck Fergus wrote about seeding his yard with box turtles. "When I see one crossing the road or sunning itself on a bank, I carry it home."

Nothing made my own kids happier than to have box turtles around the yard of our summer home. They would play with them for a while and then let them go, knowing that the turtles would be around whenever they wanted to play with them again.

THE BOX'S KIN

There are two distinct species of box turtles in North America. One is the eastern, *Terrapene carolina*, about which this chapter has been written, and its two subspecies, the Florida of Florida and the three-toed of the Southeast to Texas. The other species is the ornate box turtle, *Terrapene ornata*, sometimes called the western box turtle, of the midlands to Texas. The ornate's subspecies, the desert, lives in the deserts of the Southwest.

The other common dry-land turtle of North America is the wood turtle, *Clemmys insculpta*, found in the Northeast and upper Midwest.

. . . G.H.H.

BOX TURTLE FACTS

Description: A dry-land turtle, with a 4- to 8-inch dome-shaped shell, heavily marked with yellow, orange or olive on black or brown. Bottom of shell hinged, allowing turtle to withdraw limbs and shut tightly. Weighs about 1 pound.

Habitat: Woodlands, field edges, forest bottomlands and suburban backyards near water.

Habits: Diurnal and solitary, except when breeding. Possible homing instinct. Hibernates during winter in 2 feet of soil, rotted log or stream bank.

Form/Nest: Sleeps at night in shallow form which is dug with front feet in soil and litter. Lays eggs in flask-shaped nest, dug with hind feet at night.

Food: Young are carnivorous, adults herbivorous. Food includes earthworms, slugs, snails, insects and their larvae, crayfish, small frogs, toads, snakes and carrion. Vegetables include leaves, grasses, berries, fruits and mushrooms. Will eat lettuce and hamburger in captivity.

Voice: None. Only noise is hissing made when air escapes from shell while closing.

Locomotion: Shuffling waddle on tiptoes as legs alternately shift shell over the ground.

Life Span: May live 60 to 80 years in wild; some believed to be over 100.

THIRTEEN

OPOSSUM
Dim-witted but Remarkable

If I hadn't known better, I might have been frightened by my confrontation with the scraggy, ratlike animal that was threatening me. Its lips were drawn back in a sardonic grin to reveal its 50 needle-sharp teeth. Beady black eyes stared at me; the mouth drooled. I was certain the creature was hissing at me.

We scrutinized one another for a moment, both recovering from the first startling awareness of the other. Then I continued down the driveway and the opossum waddled off into the blackness.

This opossum, like most of its kin, was not pugnacious. Despite first impressions, the opossum is generally a passive animal. Baring of teeth makes it look ferocious, but if push comes to shove, the opossum will probably turn tail and scramble away, as this one did, or even faint from fright, "playing 'possum."

HOMELY BUT DISTINCTIVE

The opossum is unique among North American mammals . . . one of a kind. Its quaint natural history makes it a favorite among many backyard wildlife enthusiasts who have empathy for this gentle

The gentle opossum is unique among North American mammals.

Opossum tracks are star-shaped, like asterisks.

creature that has no chance of winning a wildlife beauty contest.

The average adult opossum is the size and weight of a household tabby—about 24 to 26 inches long, including tail, and usually 6 to 12 pounds. It has jet-black eyes, a 12-inch tail, pink nose, naked black ears, a sharply pointed snout, and a mouth that is chock-a-block with those 50 teeth—more teeth than are found in any other North American mammal.

Its long coat can look fine and well groomed at times; at others, it looks windblown and unkempt. The underfur is white with black tips; the guard hairs are white. Overall, this gives the animal a grizzled gray appearance, but I've often seen opossums crossing our patio at night that seemed to have a beautiful silvery luster to their pelage. Yet the opossum cannot by any stretch of the imagination be labeled a "pretty" animal.

For its size, the opossum's legs are rather short. Each of its pale pinkish feet has five toes. The first toe on the hind foot is very much like our thumb. Unlike the other toes, it has no nail, and it is opposable. That is, it can meet the other toes to grasp, as our thumb is able to touch all the other fingers on our hand. Feet like this make the opossum's tracks in snow or mud very distinctive. The tracks are star-shaped, like asterisks. Between the footprints is another continuous track. This is made by the tip of the tail, which the animal characteristically carries low, with tip downward.

Something else makes the opossum distinctive. It is North America's only marsupial. Like the kangaroos, koalas, wombats and wallabies of Australia, the opossum female has a pouch, or marsupium, in which she nurtures her young until they are weaned.

The opossum's eyesight is most kindly described as myopic. However, its senses of smell and touch are acute, and its hearing is quite good except in some of the lower ranges. I believe the only reason the opossum and I met on that October night was that there was a brisk wind blowing my scent away from the opossum and rustling the dry autumn leaves in the apple and maple trees to cover the sound of my approach.

"IT'S LIKE NOTHING BUT ITSELF"

The first time American naturalist Ernest Thompson Seton saw an opossum was when a friend brought a dead specimen to him. Seton said, "I never saw one alive. I don't know whether it's an active, alert thing like a squirrel, leaping from bough to bough, or a slow, cautious thing like a coon." The supplier of the opossum replied, "It's just the slowest, stupidest, crawlingest thing that ever was in the woods, and it's like nothing but itself."

Moving both legs on the same side in unison, it has a slow, awkward and plodding gait. Even at full speed, an opossum rarely tops 8 mph, making it no match for a dog in hot pursuit. Unlike the dog, however, the 'possum can scramble up a tree to escape.

It feels quite at home in trees, and maneuvers very well among the branches with its grasping hind feet and prehensile tail. The tail is not as strong and useful as those of the South American monkeys that hang and swing from their tails, but if the opossum is in a precarious position, the tail helps by acting like an anchor. The 'possum can wrap it around a small limb for extra security, for example, while trying to reach a particularly appealing persimmon that's just barely within reach.

The tail is not strong enough to hold the full weight of the adult animal for more than a very short time. And regardless of those charming folk tales that have been retold through the years, opossum youngsters do not hang by their tails from their mother's tail on nightly family sorties. "This idea was first foisted on the world by a picture done in 1717, to illustrate the Surinam opossum," according to Doutt, Heppenstall and Guilday, authors of *Mammals of Pennsylvania*. "Since that time, one naturalist after another has perpetuated the fallacy, using modifications and improvements on the original drawing. One naturalist actually used a photograph to illustrate this method of accommodating the young, but under investigation it was found that the animals involved were stuffed and placed in the position desired! It has only been recently that anyone has bothered to ascertain that an opossum cannot carry her tail arched over her back!"

NO SOCIAL LIFE

Solitary, except for females with young or for very brief interludes during the breeding season, opossums do not have an organized social structure. Only rarely have two adults been seen together in the wild. If two males encounter one another, they will fight viciously. Usually, they can detect the other's presence and merely avoid confrontation.

The opossum is slow and plodding, but it can sometimes escape a pursuer by climbing a tree.

Perhaps this is why opossums don't have many vocalizations; they don't have much need to communicate. They can hiss and growl and screech, and during courtship, the male makes a clicking noise to the female. Females also make this sound to youngsters.

PASSIVE SIMPLETON

Generally, opossums just amble placidly through life. Nonaggressive, slow and dim-witted, they are the pacifists of the animal world.

Typical of their passive nature was an incident near Washington, D.C., related by Devereux Butcher to Fred M. Packard: "Approaching a chestnut snag in the woods last spring, I looked up and saw three full-grown 'possums peering down. . . . Remaining very still they gave me the impression that they were trying to escape notice. But the 'possums were destined for trouble anyway, and this presently came to them in the form of a scolding titmouse," he reported. "Flitting about, this titmouse became very bold, always coming closer. . . . Soon the bird alighted for an instant on an outstretched tail. Unthreatened, he eventually alighted on the animal's rump, pecking vigorously—or so I thought—and causing the 'possum to jump a little with pain. After a few moments I discovered that what was

Nonaggressive, slow and dim-witted, opossums just amble placidly through life.

happening was that the titmouse was pulling out hair for nest lining. Upon acquiring a beakful, the bird disappeared. In a short while he returned for more . . . so bold as to stand on the 'possum's back and uninterruptedly pull out hair for all he was worth."

Submissiveness or stupidity? In this case it could have been either or both. "As to intelligence, most observers agree that the opossum has none," insisted Seton. "It is a silly grinning idiot."

If one assumes that brain size is a measure of intelligence, the opossum emerges at the bottom of the IQ range for an animal with its skull size. Vernon Bailey, a former naturalist with the U.S. Biological Survey, decided to measure the brain capacity of a number of common animals by filling the brain pans with beans. The number of beans necessary to fill a 'possum's brain chamber was 21. In comparison, those cited in the following list make the opossum look imbecilic indeed: skunk, 35 beans; porcupine, 70; raccoon, 150; red fox, 198. Seton conducted his own bean study and obtained similar results.

Nevertheless, opossums have enough gray matter to account for some memory. A study conducted in Cornell University's Section of Neurobiology and Behavior showed that opossums avoid mushrooms that in the past made them sick. Dr. Scott Camazine fed opossums 18 species of common mushrooms. The *Amanita muscaria* caused vomiting. Months later, when presented with a mushroom of the same species, the opossum sniffed the mushroom but would not touch it, even when obviously hungry. The study also showed that the opossums apparently remembered an unpleasant food for up to a year, even though they'd had only one experience with it.

Seton would probably have found that hard to believe. "The 'possum is the most hopeless fool of the woods," he wrote. "It will enter any kind of trap, no matter how often it has previously suffered from such contrivances; and when caught by some hard-gripping steel, it has not wits enough even to get mad; it simply looks bored, scared, and non-resistant."

IT SURVIVED THE DINOSAURS

Wits, obviously, are not the means by which the opossum has survived for over 50 million years. Saber-toothed tigers, dinosaurs and mammoths perished, but the opossum, *Didelphus marsupialis*, plodded on, a stupid, slow, passive little bumpkin. Today it survives as North America's oldest and most primitive mammal, often referred to as a living fossil. The only order of mammals in the world that are more primitive than the marsupials are the Monotremes of Australia, which include the platypus and the spiny anteaters.

DIXIE DWELLER ADOPTS NORTHLAND

Not only has the opossum survived, it has greatly expanded its range in recent times, spreading north from Central America. Firmly established long ago in Dixieland, Br'er Possum is now found throughout the eastern United States as far west as Colorado and north to Wisconsin, Minnesota and southern Ontario. Commonly known as the Virginia opossum, it has been successfully introduced into California, Oregon and Washington, and is the same species that lives as far south as Costa Rica.

Ideal habitat is a woodland, farmland or suburban neighborhood with water nearby, but the opossum is so adaptable that it thrives in nearly any habitat, from wilderness to inner city. It is a common visitor to backyards that provide its basic necessities: a water source, food (anything from garbage to birdseed to ripening garden vegetables) and appropriate den sites. Not being fussy, the opossum is easily accommodated.

Ideal habitat is a woodland, farmland or suburban backyard with water nearby.

NO OPOSSUM ROMANCE

Not even the skunk cabbage is poking through the earth when opossums begin their breeding season at the end of winter. In the extreme South, like Florida and Texas, opossums may begin mating as early as January. Farther north, it's usually February or even March in the uppermost regions of its range, like Wisconsin.

If, while on his nightly foraging expedition, a male opossum detects a female in heat, he'll follow his nose until he finds her. A female will be receptive to a male only while she is in heat.

Courtship among opossums is nearly nonexistent. The sexes encounter one another, mate, and then continue on their way, perhaps never crossing paths again.

Males are relatively submissive in the presence of a female. They often make an odd metallic clicking sound with their teeth or tongue. The purpose of this isn't really known, but it's probably an attempt to reassure the female and prevent her from treating the male hostilely. Researchers who have kept captive opossums report that the clicking is heard continuously during the night when a male is in the presence of a receptive female.

Some have reported that the male does a little "dance," too. With head and forelegs up, hindquarters down and tail extended straight

Opossums begin their breeding season at the end of winter.

outward, the male jerks forward while the tail flips back and forth. It's certainly not as impressive as the backward courtship somersault of a bird of paradise, nor the death-defying aerial dives of golden eagle pairs locked together by their talons, but it usually does the trick for opossums.

The courting male seems obsessed with nuzzling the female's hindquarters with his snout, and finally mounts her from the rear. He uses his mouth to grasp the fur on her nape and embraces her around the midsection with his forefeet. J. J. McManus was able to observe captive opossums mating and gave the following account in *The American Midland Naturalist:* "Positioned thus atop the female, the male shifts his weight laterally, causing the pair to topple over on their sides. Reynolds (1952) suggested that fertilization is successful only when the pair lies on the right side. Of the two matings I observed, one was on the right side, the other on the left. Only the female which copulated on the right side was fertilized." The actual copulation continues for 15 or 20 minutes, according to McManus.

McManus believes that the female will accept a male only once during the estrous period. However, if a pregnancy does not result during the 36 hours or so that she is in heat, she will continue to come into estrus in 28-day cycles until a breeding is successful.

None of this is terribly unusual, and it's not nearly as colorful as the myth that still survives today in some localities about the way in which opossums reproduce.

This particular folk tale would have us believe that the male impregnates the female though her nose, and later she blows or sneezes the tiny babies into her pouch. "The nostril idea originated from the fact that the male reproductive organ is two-forked and the only double opening visible in the female is the two nostrils," explains Hartley H. T. Jackson, author of *Mammals of Wisconsin.* What the old-timers who authoritatively passed along this myth couldn't easily see is that the female's vagina is also forked.

If the female becomes pregnant as a result of the breeding, her young will be born a mere 12½ days later, the shortest gestation of any North American mammal.

ANY OLD PLACE IS HOME

Opossums are transient, often using whatever den is most convenient within their somewhat loose home range, rather than returning to a particular den each day. Females with young tend to be the exception to this, sometimes using the same den site for weeks at a time. As with their other basic needs, opossums are not particular about what makes a suitable den. Often it is an abandoned woodchuck burrow, a hollow tree or log, a brush pile or a woodpile, but

Opossums often use whatever den is most convenient.

The prehensile tail is used to gather and transport materials for nests.

they're just as likely to hole up in a rocky crevice, a culvert, a barn, a drainpipe, a thicket or under a porch.

The nest itself is made of leaves, and sometimes papers or rags are collected and brought into the den. For this, the prehensile tail comes into use. Perhaps the best description of how both male and female opossums use the tail to gather and transport materials for their nests was given by Luther Smith in the *Journal of Wildlife Management.* On an early-morning excursion in the Missouri woods, Smith's attention was attracted by a rustling in nearby dry leaves. "The noise was made by a young opossum slightly more than half grown," he wrote. "The animal came out of a hole in the ground about eight feet from where I stood, and proceeded to select small mouthfuls of two or three leaves each." The leaves were taken out of the mouth with the forepaws and passed back to a position in front of the thighs. The hind feet slid them along into a loop in the tail.

Smith claimed that six or eight mouthfuls of leaves filled the tail loop. "The action was rapid and the leaves were in almost continuous motion from the time they were picked up from the ground until they came to rest in the coil of the tail," Smith wrote. "After the loop was filled, the opossum chose a last mouthful and, with its tail extended almost horizontally except for the loop which held the bundle of leaves, proceeded into the hole in the ground." The opossum made four such trips in about ten minutes, he reported.

SPEEDY DELIVERY

The arrival of the tiny new opossums is preceded by some restlessness on the part of the mother. Instinct drives her to begin an incessant licking process. Sitting in a fairly upright position, she thoroughly licks the furry lining of her abdominal pouch, which will serve as an incubator for her newborns. She also licks her belly fur, from the pouch to the genitals, moistening the fur in the process.

There is no difficult or prolonged labor for the opossum. In a matter of a few minutes, the female may give birth to more than 20 young. The babies are so tiny that even the first scientists to observe the procedure closely missed the wee things as they made their entrance into the world.

OPOSSUM NEWBORNS ARE PREEMIES

A newborn opossum is not a miniature version of its parent. In fact, after such a short gestation period, it is little more than an embryo. Each is less than ½ inch in total length, and typically weighs .0046 ounce. It would take more than 200 of them to weigh an ounce, and an average litter of about 16 could easily fit into a tablespoon. Like little pink grubs, they are hairless. Their eyes and ears are not yet exposed, and there is a small opening for the mouth. The back end of the grub has three nubbins which later will develop into the tail and two hind legs. The front legs, however, are fairly well developed, and come equipped with strong, sharp nails.

NEWLY-BORN OPOSSUM. ENLARGED PICTURE ON LEFT SHOWS RUDIMENTARY HIND LEGS AND WELL DEVELOPED FORE-LEGS. ON THE RIGHT —AN ACTUAL SIZE DRAWING OF NEWLY BORN YOUNG ON A PENNY.

Newborn opossums, less than ½ inch long, are like little hairless pink grubs.

As each embryonic offspring is expelled from her body, the mother licks off the chorionic sac, preventing the newborn from drowning in the fluid that surrounded it. The young opossum, blind, with only two working legs and no help whatsoever from its mother, then begins what will be its first, and one of its toughest, struggles for survival. Entirely on its own, the youngster must laboriously drag itself up to and into the marsupium 3 inches away. This distance may not seem far, but keep in mind that the creature traveling that distance is barely the size of a navy bean.

"For locomotion, the embryo employs a kind of 'overhand stroke,' as if swimming, the head swaying as far as possible to the side opposite the hand which is taking the propelling stroke," Carl G. Hartman explained in his 1920 paper "Studies in the Development of the Opossum." He observed, "With each turn of the head, the snout is touched to the mother's skin as if to test it out."

The mother's moistening of her belly fur undoubtedly helps the little one in its effort. Jim Keefe, author of *The World of the Opossum*, notes, "Young opossums placed on the dry fur of their mother are not able to climb as easily as they can when either they or the hair is slicked down by wetting. Young have been observed to recoil from the prickly dry hair when they deviated from their proper course."

It used to be a popular theory that these young opossums were operating under "negative geotropism." In other words, they instinctively moved away from the force of gravity, driving them to climb upward to the pouch. Even when a newborn opossum is removed from its normal course and placed in another section of the mother's fur, it continues to climb upward, whether or not the marsupium lies in its path. Then someone discovered that until they are 41 days old, the babies do not have functional inner ears to respond to the pull of gravity.

"Once 'negative geotropism' was ruled out as an explanation for the tiny opossum's tendency to crawl ever upward, scientists looked for some other explanation," Keefe reports. "Close observation and a little applied common sense supplied the answer. . . . The tiny opossum, when just born, is provided with fairly well developed front legs with strong claws. The rest of the creature tapers away to hind limbs that are mere undeveloped five-lobed pads. The only means of locomotion are the front legs, the rest of the body simply being dragged. Given the impulse to crawl, as the opossum certainly is, and with only the front legs working, the natural thing is for the rest of the body to fall with the force of gravity. Thus the front of the animal is always pointed up, and that's the way it travels," he explains. "The pouch lies in its path."

Some do not survive the journey to the marsupium. Those that do immediately face another challenge. Arriving inside the furry pouch, the youngsters must find a nipple. There are usually 13 nipples, arranged in the shape of a horseshoe, with one or two in the center. The newborn opossums that reach the pouch but are unable to attach themselves to a nipple perish.

The number of opossums born may be as many as 16, 18 or even more than 20, but the typical number of young in a pouch is nine. The remainder are doomed. "Even some of these unfortunates, however, held on with their mouths to a flap of skin, or to the tip of a minute tail, while several continued to move about," Hartman remarks.

Once the youngster, with its strong sucking instinct, firmly attaches itself, the nipple literally becomes its lifeline for the next two months. Almost immediately, the nipple is drawn out to twice the 1-millimeter length it was when the baby found it. As the young ones grow, the nipples continue to extend to accommodate the increasing activity of the youngsters.

LIFE IN A FUR-LINED POUCH

Like an egg tooth dropping from the bill of a newly hatched bird, the claws on the baby opossum's forefeet, which were vital in the journey to the pouch, fall off. The claws are no longer needed and would only create problems in the crowded pouch.

Opossum youngsters grow quickly, secured to their snug, furry incubator by the nipples they will not relinquish. By the age of one week, they have increased their birth weight by ten times! On about the 17th day, the tail and hind legs show some activity, and it is possible to determine which are males and which are females.

At 36 days, tiny vibrissae (whiskers) make their appearance, and at 43 days, some body hair is visible.

Sometime between the 60th and 70th day in the pouch, the youngsters' eyes finally open, they have some control over their body temperatures, and they have grown to the size of a white-footed mouse, weighing about 9/10 ounce.

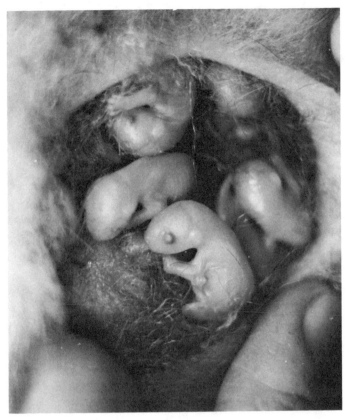

Little more than embryos, the newborns scramble into the mother's pouch and attach themselves to a nipple.

It is at this time that they begin to occasionally release their hold on the nipple. Soon they make cautious peeks out the pouch opening, and finally they venture outside.

At 75 to 85 days of age they are weaned and rarely go back into the pouch. Yet they remain with their mother until they are between three and four months old.

A LITTLE OF THIS, A BIT OF THAT

When they start to eat solid foods, they find that just about everything is edible, including a lot of items that few other creatures would consume. An opossum lives up to the term "omnivorous" as much as any animal can. Its diet includes insects, worms, beetles, ants, grasshoppers, crickets, mice, moles, bird's eggs and young, snakes, frogs,

At 75 to 85 days, the young are weaned and rarely go back into the pouch.

lizards, garbage, carrion (especially road-killed rabbits, opossums and other wildlife), corn and berries. When they are available, persimmons are irresistible. If the opportunity arises, an opossum will also wreak havoc in a hen house, which is quite a feat considering the creature's slow, clumsy ways. "It probably receives more condemnation than it deserves for its destruction of small game and birds' eggs, and its raids on chicken coops are infrequent," Hartley Jackson offers in the 'possum's defense.

Most food is eaten while the opossum sits on its hind legs, using the forepaws to hold the food. Afterward, it grooms itself, licking its paws and washing its face and body in catlike fashion.

LIFE OUTSIDE THE POUCH

The young, now too large to scramble back into the mother's pouch when threatened, find that their life-style is changing.

According to Laurence N. Gillette, who did the extensive study "Movement Patterns of Radio-Tagged Opossums in Wisconsin," once the youngsters are too large to be carried in the pouch (about 80 days after birth), they are left in the den while the mother forages for food. "The females occasionally returned to the dens during the night, but often they stayed away all night, returning just before dawn to spend the day with their young," Gillette reported. "Occasionally litters were moved to new dens. These moves usually took less than an hour. Den changes were the only times that young were observed riding on a female's sides or back."

Gillette's observation is interesting, because for years, naturalists have told us that the mother opossum trundles her out-of-the-pouch dependent young with her on her nightly forays. If Gillette's radiotelemetry trackings are typical of all opossums, it would appear that a female with youngsters gripping tightly to her fur with their claws and little tails must be on her way to a new den, not just out for a whiff of fresh night air.

According to Gillette, the young began to forage on their own between 90 and 95 days of age, but returned to share a common den with their mother during the day.

Gradually, the young ones expand their sphere of operations,

Opossum youngsters remain with their mother until they are between three and four months old.

and eventually occupy dens of their own. "Each animal appears to benefit from the discoveries of den sites made by its siblings," Gillette said. "They occupied common dens with littermates frequently for as long as three months. . . ."

MAKING ROOM FOR THE NEXT LITTER

In warmer areas, two litters seem to be common. In the northern part of its range, the opossum may have only one litter a year.

According to Gillette's study, females that have only one litter use the same den until their young no longer return (generally less than four months after birth). "Adult females that did produce second litters moved to new dens just prior to the birth of the litter, abandoning any first-litter animals that still returned. . . ." Yet, early American naturalist John Bachman wrote, "In the month of May, 1830, whilst searching for a rare species of Coleoptera (beetle), in removing with our foot some sticks composing the nest of the Florida Rat, we were startled on finding our boot unceremoniously and rudely seized by an animal which we soon ascertained was a female opossum. She had in her pouch five very small young, whilst seven others about the size of full-grown rats were detected peeping from the rubbish."

The young gradually begin to forage on their own, but for a time return to share a common den with their mother during the day.

Once on their own, young opossums disperse from the area of their birth.

It's hard to refrain from being anthropomorphic in the face of these statements. I'd like to say that opossum mothers, like human mothers, apparently differ in their parental devotion. Some don't seem to mind having a passel of young ones constantly underfoot, while others can barely wait to shoo them out the door!

Once on their own, young opossums disperse from the area of their birth. Gillette discovered that pregnant females, too, typically disperse to a new den before the birth of a new litter.

"The factors which stimulate an opossum to disperse are unknown, but there appear to be good reasons for the differences observed between sexes," Gillette reported. "The primary function of the male is to locate and mate with as many females as possible. This can be accomplished quite effectively by gradually shifting a home range."

A female might enhance her young's chances for survival by varying the locations at which her litters disperse, eliminating competition for food or den sites between her litters. "However," Gillette points out, "dispersal may serve a more important function along the northern edge of the opossum's range. Here, populations may exist only in widely scattered areas or at very low densities. Severe winters could eliminate all opossums over large areas. Under these circumstances, dispersals by females carrying young may enable the opossum to repopulate depleted areas, or expand into new areas more rapidly than would be possible if only juvenile opossums dispersed as individuals."

A LIFE ALONE

Leaving the family unit is the beginning of a life of solitude for a young opossum, except for the females that will rear young for a few months each year.

Even as juveniles in the protection of the old female, the siblings do not play. Play among wild babies helps them to develop the skills they'll use in later life. The rough-and-tumble wrestling matches that are so common among litters of other young mammals, however, is completely lacking among opossum youngsters. Perhaps it's because the opossum's social system is so undeveloped. Perhaps it's because, unlike fox kits, bobcat kittens and wolf pups, their adult life-styles will not demand that they fight to survive. The opossum, after all, is the ultimate pacifist. Except in very rare cases, it will submit rather than fight.

PLAYING 'POSSUM

The animal is not without its own defense mechanism, however. When threatened, the animal does what it is famous for—it "plays 'possum," feigning death.

Like many aspects of hibernation and migration, the 'possum's death feint still has scientists scratching their heads. Until recently, the accepted belief was that the opossum's deathlike trance was a deliberate effort. But John Burroughs aptly pointed out that a 'possum is not himself wise; "Nature has been wise for him."

Today, most believe the dead-opossum act is a completely involuntary reaction to a dangerous situation. As our artist/naturalist friend Ned Smith says, "It's the totally involuntary reaction of a beast who's scared silly and too stupid to know what to do about it." Perhaps the opossum literally faints from fright. Or it's possible that the fright causes a chemical substance to be released into the animal's system which activates the condition (if so, the substance has yet to be identified).

When threatened, the animal "plays 'possum," feigning death.

As the opossum makes its way through the night, it may be confronted by a bobcat, fox, coyote, badger, wolf, owl or domestic dog, each determined to make a meal of the slow, stupid, bumbling opossum. Of these, the dog is probably its worst enemy.

When the 'possum becomes aware of its dilemma, it has some alternatives. It can make a run for it, and may actually escape from a nonclimbing predator if it can climb a nearby tree. On the ground, an opossum has little chance of outdistancing its pursuer. "The usual gait under such circumstances is a waddling trot in which the feet go much faster than the body," Ned Smith contends. "Really pressed, the pink-soled feet become a blur and the outstretched tail stiffly describes lopsided arcs and circles in the air. Generally, the critter has a hideaway in mind—a hole or hollow log, or perhaps a climbable tree—in which he'll escape."

Or the opossum can stay and fight, an option rarely used.

If things are really grim, its eyes glaze over, its teeth are bared, the tongue lolls to the side, and the 'possum sinks to the ground, successfully simulating death. No amount of prodding, poking or shaking will revive the animal from its catatonic state. The opossum apparently is incapable of feeling even things that should cause extreme pain. It doesn't even flinch. Because many predators are not interested in the flesh of animals they have not killed themselves, they may nudge the 'possum a few times, sniff it, and then lose interest and move on. This often is the opossum's salvation.

Some time later, the dim-witted survivor regains consciousness and continues slowly on its way.

Two enemies from which the deathlike trance will not save the opossum are the automobile and man. Each year, countless opossums are killed on highways. They are attracted to the carrion of others that have already died under the wheels of speeding autos, and in the midst of their dinner, don't have the intelligence to realize the significance of the headlights rapidly descending upon them.

Nor is feigning death likely to work when being pursued by 'possum hunters. Mainly a sport of the South, 'possum hunting is typically done at night with dogs, a flashlight and a gunny sack. The dogs chase the opossum up a tree and keep it there until the hunters arrive. The men shake the tree (it's often nothing more than a sapling), and the 'possum usually falls to the ground, often in its coma-like state. Many a 'possum has been bagged alive under these circumstances, then easily dispatched. In many parts of the South, hunters insist that " 'possum meat is very, very fine" when roasted with sweet potatoes. In the rest of its range, the opossum is not hailed as being particularly tasty.

"Playing 'possum" is a merciful prelude to the end of life for this creature, whether in the jaws of a predator or by the swift bullet of the 'possum hunter. "When the predatory teeth reach his vitals, the opossum does not know it," Alan Devoe reminds us. "The flickering little light that glowed in his small brain—the hazy understanding that comprised not much more than a knowing of the good smell of maternal fur, the pungent taste of persimmon, the obscurely apprehended joy of slow swingings in high moonlit branches—has long before that time been automatically extinguished into its last oblivion."

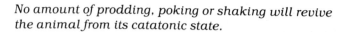

No amount of prodding, poking or shaking will revive the animal from its catatonic state.

In spite of this marvelous defense mechanism, the opossum's chances for a long life are slim. In addition to all the enemies already mentioned, opossums suffer from a number of diseases and internal parasites, probably because of their unrefined food habits. Aerial applications of chlorinated hydrocarbon insecticides, such as dieldrin, have been responsible for opossum deaths, too.

WANTED: EARMUFFS AND TAIL WARMERS

If the young opossums survive into the autumn, they probably will have grown fat like their elders. Opossums do not hibernate, but in the colder regions of their range, they wait out the extremely cold spells sleeping in their dens. Not originally a creature of the northland, opossums at this end of their range commonly suffer frostbite on their naked ears and tails. Opossums with tips of tails or ears missing from frostbite are fairly common in the northern states.

AGAINST ALL ODDS

With so many odds against them, few opossums live more than two years in the wild; most survive about one year. Yet the species itself carries on, as it has for millenniums. Certainly not because it is intelligent like a fox, or ferocious like a weasel, or plucky and agile like a chickadee. The opossum's success as a species, it seems, is entirely the result of its fecundity. With each female capable of producing as many as two dozen offspring a year, it would take more than the opossum has yet had to endure to wipe out such a prolific mammal.

EASY TO PLEASE

Any backyard that is attractive to songbirds, squirrels, cottontail rabbits and other common backyard wildlife will be attractive to opossums, too. They are fare less fussy in food and cover requirements than most of the other species in this book. A backyard that provides some shelter for a daytime den, has trees, food (birdseed, table scraps, cultivated or wild fruits) and a water source either on

Any backyard that's attractive to songbirds, squirrels, rabbits and other common backyard wildlife will probably have opossums.

the property or nearby is undoubtedly frequented by opossums. Remember that they're active at night, so you'll need to illuminate the area to observe them.

A POSSUM IS NOT A 'POSSUM

Our Virginia opossum is one of 250 species which compose the Order Marsupialia, but is unique in being the only North American representative. Marsupials are the primary mammal form on Australia, New Zealand and nearby islands, with species ranging from the tiny 2-inch-long pygmy planigale to the 150-pound red kangaroo, the largest living marsupial. Some of the more familiar Australian marsupials are the koala, the wallabies and the wombat. The sugar glider looks much like our flying squirrel, and is similarly equipped with patagia for gliding from tree to tree, but, like the other marsupials, rears its young in a pouch. Australia also has a number of *possums*, which are different from opossums.

In the New World, 70 species of the opossum family are found from southern Canada to the tip of South America. Among these are the South American mouse opossum, the water opossum, the short-tailed opossum, the colocolo and the rat opossums. None could be mistaken in appearance for the Virginia opossum.

. . . K.P.H.

OPOSSUM FACTS

Description: A housecat-sized animal, furred in grizzled gray. Beady black eyes, pink nose and feet, black ears on top of its head, pointed snout and naked, light-colored tail.

Habitat: The opossum finds that wood lots, wilderness, farmland, parks, city neighborhoods and suburban backyards suit it just fine. Ideal habitat is wooded with water nearby.

Habits: Nocturnal; rarely seen abroad in daylight. Slow-moving, solitary. Sometimes falls into a deathlike state as a defense mechanism.

Den/Nest: Hollow trees or logs; abandoned burrows of woodchucks, skunks or foxes; in culverts or drainpipes; under porches; in thickets, woodpiles or brush piles; in barns or outbuildings. Nest consists of leaves carried into the den.

Food: Omnivorous. Scavenges carrion, especially in the form of roadkills and game crippled by hunters. Eats all types of berries, persimmons when available, some corn, occasionally a henhouse chicken, sometimes wild bird eggs and nestlings, many insects, mice, moles, frogs, lizards, snakes.

Voice: Hisses, screeches and growls, but is generally silent. Male in the presence of female in estrus—and sometimes a female to youngsters—makes a metallic clicking sound.

Locomotion: A slow, plodding, waddling gait on the ground, both legs on one side moving in unison. Also at home in trees. Climbs relatively well and uses prehensile tail as an anchor line to secure positions in trees.

Life Span: Few live to the age of two years; most probably survive about a year in the wild.

The screech owl doesn't screech. Its call is wavering and tremulous.

FOURTEEN

BACKYARD BIRDS
Charming, Colorful Companions

"We had a hoary redpoll at our feeder this afternoon!" Janet and Tom Rost announced to us over the telephone. Our conversations with the Rosts often center around what has showed up at our respective bird feeders.

Tom, a renowned artist, and his wife Janet live in a small community near Lake Michigan. We live farther inland about 20 miles, and the difference in our surrounding habitats makes for interesting comparisons in the bird species we each have in residence.

For example, both white-breasted and red-breasted nuthatches are common year-round residents at the Rosts'. We see white-breasted nuthatches daily, but on the occasions when a red-breasted appears, it's special for us.

On the other hand, we've been able to cite the appearance of a mockingbird (unusual this far north), a Harris' sparrow, a white-winged crossbill and a nesting red-bellied woodpecker. They'll respond by telling us about the rose-breasted grosbeaks that spent the winter in their backyard. And so goes the continuing Harrison-Rost game of bird-sighting one-upmanship.

We find that comparing notes with other backyard bird watchers

Spectacular black-and-orange Blackburnian warblers are often spotted in backyards during spring migration.

is fairly typical among friends with whom we have this common interest.

No matter in what part of the country we live, regardless of season, there are birds to be seen in our backyards. They are the most visible and most constant form of wildlife in our gardens. They are colorful, active, easy to observe, relatively tame, fascinating to get to know and easy to attract. The three necessary elements are cover, food and water.

Usually there is a change in the cast of bird characters from summer to winter. Because many species migrate, the birds that court, build their nests and raise young in our backyards are not always the same species that visit our feeders daily throughout the winter. This, too, adds excitement to the sport of bird watching.

Most bird watchers get involved in their sport by observing and learning to identify the birds they see in their own backyards. Nearly everyone is already acquainted with robins and cardinals. They are often the first birds we learn to recognize as children. Then, the backyard bird watcher is captivated by the antics of a little gray-and-black ball of fluff and learns that it is a chickadee. Orioles, hummingbirds, various sparrows, grosbeaks and finches are also learned quickly.

Then, one day in spring, a brilliant flash of color is seen in the top of a backyard tree. The binoculars are grabbed, the field guide is checked, and a Blackburnian warbler is identified. The next day, a male scarlet tanager is spotted in the same shrub as a male indigo bunting. Lists are started and compared with fellow bird enthusiasts. Another bird watcher is hooked!

In the years that George and I have maintained a list of the birds we've seen or heard on our ½-acre property, we have recorded 165 different species. We believe that you'll find, as we did, that getting to know your backyard bird visitors enhances your enjoyment of them. It may lead to a sport that could lure you off on exotic tours to places like the Yucatan Peninsula, the South Pacific or Africa to add a great many rarities to your birding life list.

MOURNING DOVE

From the Atlantic to the Pacific, Mexico to southern Canada, the mourning dove, *Zenaida macroura*, is among the most popular backyard residents. In early spring, before the last of the snow has fallen, our resident mated pair build a flimsy nest of twigs lined with a few grasses in a nearby spruce. Two or three pure-white eggs are laid. When the young hatch, they are fed a fluid called "pigeon milk." Each youngster pushes its bill inside the bill of the adult while the parent regurgitates the pigeon milk into the mouth of its offspring. Each pair of mourning doves may raise from two to four broods a year.

From coast to coast, the mourning dove is among the most popular backyard residents.

Doves visit our bird feeders all year long. They seem to feel most comfortable eating cracked corn from a platform feeder or on the ground, but doves in our area have taken to eating niger seed out of a hanging tubular feeder. Water is also very important to doves and should be provided.

SCREECH OWL

The eastern and the western screech owls are look-alikes, except in color. The eastern, *Otus asio,* may be either gray or red (rusty brown); the western, *Otus kennicottii,* is brown.

The nocturnal screech owl doesn't actually screech. Its call is a wavering, tremulous cry that some feel has an eerie, ghostly sound. Only 8 to 10 inches long, it is our smallest owl and has tufts or "horns" on its head.

A year-round resident from coast to coast, with the exception of the extremely high mountain areas of the Rockies, the screech owl often begins nesting when there is still snow on the ground. It lays four to five white eggs in a tree cavity or man-made nesting box. Baby screech owls stay in the nest until they are about a month old; then their parents teach them to hunt for their own food.

Screech owls are usually carnivorous, feeding heavily on mice, voles and other small rodents, insects, reptiles and occasionally a small bird. Therefore, they aren't interested in the typical feeding-station offerings. However, they will be attracted to water. On New Year's Eve a few years ago, a screech owl came to our heated, recirculating pond at the stroke of midnight for a drink. We thought that was a very good omen for the New Year!

HUMMINGBIRDS

In the East, our only hummingbird is the ruby-throated, *Archilochus colubris.* In the West, especially the Southwest, however, there are over a dozen, with the broad-tailed, *Selasphorus platycercus,* and the black-chinned, *Archilochus alexandri,* usually the most common.

Hummingbirds are iridescent feathered jewels that can usually be attracted to hummingbird feeders filled with sugar water. On an annual week-long trip to northern Wisconsin, we pack a humming-

In the West, the black-chinned is one of the most common humming-birds.

bird feeder, sugar and red food color. When we arrive at the cabin, we fill the feeder with red-colored sugar water and hang it outside the living-room window. Within a matter of hours, the local hummers are steadily visiting the feeder.

After a courtship that involves a spectacular pendulum flight by the multicolored male, the less colorful female builds a dainty nest of lichens, cottony plant fibers and spiderwebs on a horizontal tree limb. The white eggs are the size of navy beans.

The hummingbird's power of flight is incredible. With a wing-spread of about 4 inches, the hummingbird can fly a mile a minute —up, down, forward or backward. In its southward autumn migration, the ruby-throated flies a nonstop 500-mile marathon across the Gulf of Mexico.

WOODPECKERS

A backyard feeding station that offers beef suet is bound to be visited year-round by one or more species of woodpeckers. We especially enjoy watching the adult downy woodpeckers teach their fluffy youngsters to eat the suet from the feeder on our ash tree.

The little downy, *Picoides pubescens,* and the slightly larger hairy woodpecker, *Picoides villosus,* are perhaps the most likely to be seen at feeders, but a number of others also routinely visit back-yards.

A backyard feeding station that offers beef suet is bound to be visited year-round by woodpeckers like the downy.

Among these is the northern flicker, *Colaptes auratus*, a nationwide resident. In the East, its wing shafts are yellow; in the West they are red.

The yellow-bellied sapsucker, *Sphyrapicus varius*, and the red-headed, *Melanerpes erythrocephalus*, and red-bellied, *Melanerpes carolinus*, woodpeckers, are spectacular birds that are common throughout the eastern states.

Woodpeckers nest in tree cavities, which they excavate themselves. Their eggs, usually three to six, are pure white.

In the summer, young downies and hairies may accompany their parents to the suet feeder.

FLYCATCHERS
(Wood-pewee and Phoebe)

From late spring through midsummer, we hear the plaintive but persistent *pee-o-wee* whistle of the eastern wood-pewee, *Contopus virens*, high in the oaks on the wooded side of our driveway. The song of the western wood-pewee, *Contopus sordidulus*, is quite different, described by Roger Tory Peterson as "a descending burry call."

The eastern and western wood-pewees are buff-colored with two white wing bars on their dark wings.

Built on a horizontal tree limb, their shallow nest cup of grass, plant fibers and spiderwebs or moss holds three creamy white eggs.

Wood-pewees are summer residents only, wintering in Central and South America.

The eastern phoebe, *Sayornis phoebe*, common east of the Rockies, and the black phoebe, *Sayornis nigricans*, and Say's phoebe, *Sayornis saya*, of the West are members of the flycatcher family.

Grabbing their insect food on the wing, phoebes, like most flycatchers, are rather nondescript little birds. An exception is the black phoebe, which is charcoal-black except for its white breast and belly.

The eastern and black phoebes build nests of grasses, weeds and plant fibers mixed with mud, lined with finer grasses, hair and wool, and plastered with mud to walls, under bridges, or on ledges.

The Say's phoebe does not use mud in its flat nest of grasses, moss, wool and hair.

A phoebe's three to five eggs are white, and there may be small dustlike spots on one or two of them.

Not attracted to the offerings of seed and suet at most feeding stations, phoebes and wood-pewees might visit backyard water areas.

The eastern phoebe is a member of the flycatcher family.

SWALLOWS AND MARTINS

When we see families of tree swallows congregating on the telephone wires beyond our garage, it reminds us that the days of summer are numbered. They are forming flocks before migrating back to their winter homes.

Tree swallows, *Tachycineta bicolor,* barn swallows, *Hirundo rustica,* and purple martins, *Progne subis,* enjoy a special place in the hearts of many backyard birders. They have a well-deserved reputation for being voracious insect eaters. While they won't eliminate your mosquito population, they'll certainly do their part in helping to control it.

Barn swallows plaster their adobe nests to beams, eaves and walls. Tree swallows and purple martins readily accept man-made birdhouses, especially if they are near water. The four or five eggs of these species are pure white.

Unless there have traditionally been martins nesting nearby, you may have trouble attracting them. They seem to go back to the same neighborhoods year after year. Many backyard wildlifers wait until they see the first purple martin "scouts" returning in early spring before putting apartment houses out on high poles. This minimizes the chances that the martin houses will become starling or house sparrow ghettos.

JAYS

Jays are boisterous, gregarious, bold, colorful birds that frequent backyards and backyard bird feeders. They are omnivorous, and at feeding stations eat seeds as well as suet and peanut butter. The blue jays in our neighborhood have become adept at imitating the high-pitched call of the red-tailed hawk, which they use when they're approaching our bird feeders. By the time they arrive, the birds that had been feeding there have scattered in panic in response to the sound of a predator.

The blue jay, *Cyanocitta cristata,* is the common jay east of the Rockies. In the West, the Steller's jay, *Cyanocitta stelleri,* and the crestless scrub jay, *Aphelocoma coerulescens,* are found in and west of the Rockies. The scrub jay also occurs in the Florida peninsula. In the extreme northern United States and throughout the Northwest,

Purple martins enjoy a well-deserved reputation for being voracious insect eaters.

the gray jay, *Perisoreus canadensis*, also known as the whiskey jack, is a relatively tame member of the jay family.

Blue jays often nest near houses, sometimes in trellises or in vines growing on porches or exterior walls. Jay nests are bulky masses of bark, twigs and leaves, built by both sexes. In most jays, the incubation of the three to five eggs is mostly or entirely the task of the female.

CHICKADEES

Whenever we put out a new bird feeder, no matter what style or where it is placed, we can count on the black-capped chickadees being the first customers.

The black-capped, *Parus atricapillus*, is the common chickadee throughout most of the country. In the Southeast, the Carolina chickadee, *Parus carolinensis*, is a fairly common family member, and in the West, the mountain chickadee, *Parus gambeli*, fills this niche in coniferous habitats.

Chickadees are the darlings of the feeding station. Their acrobatics, general bright-eyed perkiness and constant *dee-dee-dee-dee* conversation among themselves make this diminutive bird a favorite wherever it is found.

They are year-round residents within their range. During the winter, small bands of chickadees travel together, wandering over

Acrobatics, bright-eyed perkiness and constant conversation among themselves make the black-capped chickadee a feeding-station favorite.

small areas from woods to fields to backyard sunflower seed and beef suet feeders.

Chickadees nest in rotting trees or stumps where they can manage to peck out a cavity with their small bills. They will occasionally use a birdhouse with an opening of 1⅛ to 1⅜ inches. Their five to eight delicate white eggs are speckled with reddish brown.

TUFTED TITMOUSE

A bright-eyed imp of the eastern United States, the tufted titmouse, *Parus bicolor*, is most common throughout the Southeast. Though rather plainly colored, it is easy to identify because no other mouse-gray bird in the East has a crest. Its clear, loud, *pee-do, pee-do, pee-do* ringing through February days adds a touch of spring to our days in late winter.

The tufted titmouse's life-style is much like that of its close relative, the black-capped chickadee.

WHITE-BREASTED NUTHATCH

When we hear a nasal *yank, yank, yank*, the upside-down bird is nearby. Nuthatches are the only backyard birds that typically go down tree trunks headfirst.

Through the year, all across America except for a strip that runs from western Texas northward, this little bluish-gray, black-capped bird is a common backyard resident.

Nesting in an abandoned woodpecker hole or natural tree cavity that it lines with grasses, fur, twigs and bits of bark, the white-breasted nuthatch, *Sitta carolinensis*, lays about eight white eggs.

The red-breasted nuthatch, *Sitta canadensis*, of the North and the brown-headed nuthatch, *Sitta pusilla*, of the Southeast are frequent backyard visitors in those regions.

Nuthatches readily use bird feeders, eating both beef suet and sunflower seeds.

WRENS

The first day that we hear the male house wren singing his boisterous, bubbling song from our apple tree in spring is always a happy one for us.

Like other male wrens, ours spends the next week or two stuffing all possible nesting sites with crude dummy nests of twigs and other materials. This includes the other birdhouses in the vicinity, watering cans, mailboxes, and any other niche or cavity that seems appro-

The house wren is one of the easiest birds to attract to backyard birdhouses.

priate to him. When his mate joins him a fortnight later, he takes her on an inspection tour of all the available properties. She decides which she prefers, and then rebuilds the nest to her specifications, usually with fine grasses, feathers, plant fibers and hair.

The female alone incubates the six or seven white eggs, which are thickly speckled with cinnamon-colored dots. When the young hatch, her mate helps keep the hungry offspring fed on a diet of insects.

Throughout most of the United States, the house wren, *Troglodytes aedon*, is a summer bird. In the extreme South, it is a winter resident.

The Carolina wren, *Thryothorus ludovicianus*, known for its loud *tea-kettle, tea-kettle, tea-kettle* song, is common year-round in most of the Southeast. It, too, often nests in backyard birdhouses.

MIMIC THRUSHES
(Catbird, Mockingbird and Brown Thrasher)

Every day last summer, we were greeted with a *meow* on our morning walks. The gray catbird, *Dumetella carolinensis*, that was mewing to us was always in the same tree as we passed. When autumn neared and the bird flew south with the rest of its kin, we were very aware of its absence.

No other American bird is slate-gray with a black cap and chestnut undertail coverts. Those who have catbirds in their gardens are familiar with its characteristic saucy tail flicking.

Catbirds nest throughout most of the United States, except in the extreme West and extreme Southwest. Their bulky twig and leaf nests are built in dense thickets and shrubs. Four greenish-blue eggs are incubated by the female alone.

The northern mockingbird, *Mimus polyglottos*, originally a southern species, has extended its range steadily northward. It is now found throughout the year from Florida to as far north as Wisconsin and Maine, and from the East Coast to the West Coast.

The mockingbird is an accomplished mimic, singing the imitated songs of other birds in addition to its own. Its habit of repeating phrases, usually three or four times, is a good clue that you're listening to a mockingbird, and not the real owner of the song you're

The three basic elements of any successful backyard wildlife habitat are food, cover and water. This one has all three.

hearing. Unlike most songbirds, the mockingbird often sings at night as well as during the day.

It commonly nests in shrubs, vines and trees in backyards, parks and farmlands. Its four bluish-green eggs with brown blotches are laid in a bulky nest of twigs and leaves.

The song of the brown thrasher, *Toxostoma rufum*, is a series of phrases, with each phrase typically sung twice before the bird goes on to the next.

This handsome reddish-brown bird with a long tail and striped breast is the only thrasher east of the Rockies. It nests from the Gulf states to southern Canada, wintering in the southern United States.

One of the traditional signs of spring is the sighting of the first returning robins.

It lays its pale bluish-white eggs, usually four, in a nest on the ground under a shrub or hedge. Or it may build its large, bulky twig and leaf nest in a shrub, tree or vine.

The brown thrasher is a fairly large bird, about 10 inches long, but is able to conceal itself well, easily dodging in and out of thickets.

Not interested in the usual feeding-station offerings, catbirds, mockingbirds and thrashers will frequently use backyard water areas.

ROBIN

Several robin pairs nest within sight of our house each year, but it's very special when a pair chooses to nest on top of the light fixture at our back door as they have for the last two years.

The American robin, *Turdus migratorius*, is the quintessential backyard bird. In early March, one of the traditional signs of spring is the sighting of the first returning robins. In many areas, there is still a snow cover when the robins reach their breeding grounds.

By April, they are involved in their first nesting. The female constructs a cup nest, mostly of grasses and mud, which she molds to her body. In this she lays her four beautiful "robin's-egg blue" eggs.

Young robins, like other young thrushes, have spotted breasts.

Eastern bluebirds may nest in bluebird boxes that are placed on fence posts or backyard fruit trees.

Later, they become the typical robin red. Robins usually raise two broods a year, sometimes three.

Adults and youngsters alike love to bathe, and if there is a pond or birdbath in your backyard, there's sure to be a queue of robins on hot summer days. We once had nine at one time at ours!

The American robin occurs at times throughout the entire United States and Canada, as far north as Alaska.

BLUEBIRDS

On a visit a few years ago to the farm that George's grandfather owned 30 years ago, we were delighted to find that bluebirds were still nesting in the same sites that their ancestors had used when George was a child.

The eastern bluebird, *Sialia sialis*, found throughout the Eastern United States to the Rockies, is bright blue with a rusty-red throat and breast. The two western species are found from the Rockies to the West Coast. The mountain bluebird, *Sialia currucoides*, is cerulean blue, with the breast somewhat lighter than the back. The western bluebird, *Sialia mexicana*, resembles the eastern bluebird, but has a blue throat above its rusty breast.

After a sharp decline in their populations, the insectivorous bluebirds are now making a slow comeback since the ban on the use of DDT in this country.

Bluebirds are cavity nesters that must compete with the more aggressive starlings and house sparrows. They'll readily accept bluebird houses that are suitably located on posts along fence lines and in open woods, farmlands and backyards. They also nest in natural cavities, such as old woodpecker holes. In this the female builds a loose grass cup in which she lays four or five pale bluish-white eggs.

CEDAR WAXWING

These delicately colored crested beauties are usually seen in flocks feeding on fruits and berries in trees and shrubs. They also catch insects on the wing. Their thin, lispy, wheezing *zeeeee* call is a sound that we hear every morning from May to October as we walk our mile-long route.

Cedar waxwings, *Bombycilla cedrorum*, winter throughout most of the United States and into Mexico. In summer, they breed from Maine to Washington, south to northern Georgia. Waxwing pairs usually build their loosely woven grass and twig nests on the limb of a shade tree. The three or four eggs are bluish-gray spotted with brown.

Bohemian waxwings, *Bombycilla garullus*, locally common in parts of the Northwest, are somewhat larger and grayer than the cedar waxwings and have rusty undertail coverts.

The handsome red-winged blackbird males arrive on the nesting grounds a week or two before the drab, streaky-brown females.

YELLOW WARBLER

Yellow warblers, *Dendroica petechia*, aren't usually interested in what we have to offer at our summer feeding station, but they come to our recirculating pond several times a day to drink and bathe.

Of the many wood warblers that are found in North America, the yellow is surely one of the most common and easily identified.

A summer resident throughout most of Canada and the United States, the yellow warbler arrives on its breeding grounds in April. By May it is nesting, and by June the young have fledged. At the end of July, many yellow warblers are already migrating south. By the end of August, the yellow warbler has virtually disappeared from its summer breeding range, well on its ways to its winter home in Central and South America.

Yellow warblers often nest in colonies. One of the highest densities of yellow warbler nests we've ever seen was in the multiflora roses that border a parking lot at the Pymatuning waterfowl refuge in Pennsylvania. We found an active nest every few feet.

The female builds a strong, tightly woven cup of milkweed fibers, grasses and plant down in which she lays four or five whitish eggs marked with brown splashes.

RED-WINGED BLACKBIRD

The red-winged blackbird, *Agelaius phoeniceus*, is one of the most common birds in North America. In much of the United States, it is abundant year-round, but in the more northern states and in Canada, it is strictly a summer resident.

Not everyone's favorite, perhaps, but the redwing is surely as much a harbinger of spring as the robin, at least in our backyard. Inevitably, we spot our first robin of the year on the same day we hear the first redwing sing. Its joyous *cong-ar-ree* is indeed a welcome sound to those of us weary of winter.

The handsome black-bodied, red-winged males arrive first on the nesting grounds, followed a week or two later by the drab, streaky-brown females. Most of the males are polygamous, each having two or three females in his harem.

The female builds a nest, often in cattails, reeds or bushes, of rushes, grasses, sedges and mosses, which she lines with fine grasses. Her three or four pale bluish-green eggs are marbled with brown, black and purple.

Though we rarely see female redwings at our bird feeders, the males frequently zip in for a few seeds when they feel they can spare a few moments from defending their territories.

NORTHERN ORIOLE

The northern oriole, *Icterus galbula* (formerly the Baltimore oriole and the Bullock's oriole), is one of the most colorful songbirds at our feeders. In spring, when we hear the first oriole singing its song in a minor key, we impale an orange half, cut side up, on a nail or twig in a tree on our patio. Or we simply place it with cut side up on a feeding platform or table outside our window where we can easily observe it. Within a few hours, we have orioles eagerly eating it. The fruit and the bird's breast are the same vivid color!

An oriole's nest is an intricately woven pouch hanging from a tree branch. It is built by the female from plant fibers, hair, grasses and, in the South, Spanish moss. The four whitish eggs are marked with black and brown scrawls.

In the West, a number of other orioles are fairly abundant, and can often be enticed to take sugar water at backyard feeders.

TANAGERS

The male scarlet tanager, *Piranga olivacea*, is one of the most striking birds in North America. Its flame-scarlet body is accented with jet-black wings. Seen in a flowering shrub on a sunny day, it is a sight to thrill any bird watcher, no matter how jaded. In autumn, the brilliantly colored male molts into his winter garb, which is the same as his mate's year-round plumage. Where the male was scarlet in summer, he is dull green and yellow in winter.

Scarlet tanagers nest from Oklahoma and Virginia into Canada. Their flimsy twig nests holds three to five pale-blue eggs that are spotted with brown.

The western tanager, *Piranga ludoviciana*, is quite common in coniferous and aspen forests of the West. The bright yellow male has a red head, black tail and black wings with wing bars. The female is olive above and dull yellow below, much like the male in winter.

The western tanager breeds from Mexico to Washington and Montana, and into Canada. Its three to five bluish-green, spotted eggs are laid in a shallow twig nest, usually in a pine or fir tree.

Brilliant scarlet tanagers often nest in wooded back-yards, especially those with oaks.

CARDINAL

The first birds to arrive at our feeders as daylight nears and the last to visit before dark are the cardinals. When the male is courting the female in spring, we sometimes see him pass a sunflower seed from his bill to hers.

The northern cardinal, *Cardinalis cardinalis,* is unquestionably one of the most popular American backyard birds. Easily identified, even by young children, it is the only all-red bird with a crest.

The cardinal is a year-round resident throughout its range, which is primarily east of the Rockies. Even on frigid February days, we sometimes hear the cardinal's *birdy-birdy-birdy* song.

Cardinal pairs stay loosely together throughout the winter, making daily visits to the local sunflower seed feeders. In spring, they strengthen their pair bond as the breeding season approaches.

The female, occasionally assisted by her mate, builds a loose nest of twigs, leaves and bits of bark and grasses in dense shrubbery, thickets or small trees. She alone usually incubates the three or four whitish eggs that are marked with blotches of brown, gray or purple. Often, the male brings foods to her while she is on the nest. When the young hatch, he assists in feeding them.

Cardinals typically raise two or three broods a year.

The rose-breasted grosbeak's large, conical bill is well suited to cracking seeds.

GROSBEAKS

No matter what we're involved in, all work comes to an abrupt halt when we see a male rose-breasted grosbeak fly to our pond for a summer bath. A spectacular-looking bird, the male rose-breasted grosbeak, *Pheucticus ludovicianus*, has a jet-black head, back and tail, a white rump and belly, and a bright, rosy-red breast. The streaked brown female, on the other hand, looks like an overgrown sparrow.

Its northern breeding grounds range throughout much of Canada, and from the central United States to the East Coast. Usually raising only one brood a year, occasionally two, the rose-breasted grosbeak often builds its nest in the fork of a tree or shrub. It is a flimsy structure of twigs and dried weed stalks which contains four pale greenish-blue eggs spotted with browns and purples. The male and female share the task of incubation, and both sexes sing while sitting on eggs. Roger Tory Peterson describes the rose-breasted grosbeak's song as resembling a robin's, but mellower, with more feeling, "as if a robin has taken voice lessons."

Winters are spent in the West Indies and from Mexico to South America.

The evening grosbeak, *Coccothraustes vespertinus*, breeds in Canada, the far West, and the extreme northern states in the East. Most of us see this striking, large yellow bird with black wings and tail and white wing patches only in winter. Roving bands of evening grosbeaks sometimes invade feeding stations and in a matter of hours can deplete the sunflower seed supply. We've never known a backyard wildlifer to be upset by this. Those whose feeders are frequented by evening grosbeaks are the envy of fellow backyard birders.

BUNTINGS

Buntings are small, colorful birds that are usually found in thickets, hedgerows and woodland edges.

We find that they enjoy brief visits to our water areas during the nesting season and throughout the summer.

The indigo bunting, *Passerina cyanea*, a summer resident east of the Rockies, spends its winters in Central America. The male is feathered in a brilliant, almost iridescent blue. The female is quite a contrast. She is a very plain brown bird.

From the Rockies to the West Coast, the Lazuli bunting, *Passerina amoena*, is the indigo's counterpart. The male Lazuli is blue, with two white wing bars and a pale cinnamon-colored breast. The female, like the female indigo, is plain brown, but has a faint wash of blue in its wings and tail.

In parts of the extreme South, the painted bunting, *Passerina ciris*, is locally common, and backyard wildlifers in these areas often see them at their bird feeders. The male's colors seem too bright to be believable, almost gaudy. His head is rich royal blue, his underparts and rump are vivid red, and his back is yellow-green. The female is entirely yellow-green.

Female buntings weave nests of dried grasses, lay three or four eggs and incubate without help from the male. Normally two broods are raised each summer. Winters are spent from Mexico to Panama.

FINCHES

Purple finches, house finches, Cassin's finches, redpolls, pine siskins and goldfinches are the common backyard finches. Our well-stocked niger seed feeders often attract small flocks of these sweet songsters.

The male purple, Cassin's and house finches are quite similar in appearance. They are often described as resembling sparrows dipped

Pine siskins seem reluctant to give up a position on the sunflower seed feeder.

in raspberry juice because of their rosy-red plumages. Their mates are brown and heavily streaked.

Purple finches, *Carpodacus purpureus,* spend the breeding season in coniferous woodlands of the northern United States and Canada. But in winter, they are common at backyard feeding stations across most of the United States. The Cassin's finch, *Carpodacus cassinii,* spends the entire year in most of the western states, and the house finch, *Carpodacus mexicanus,* once a species that occurred westward from the Rockies, is now strongly established in the Northeast as well after having been introduced in about 1940.

Common redpolls, *Carduelis flammea,* brown-streaked with bright-red caps on their foreheads, are birds of the far north, and even during the winter, these tundra breeders don't get much farther south than Illinois and Pennsylvania. We find they are easily attracted to bird feeders, where they especially enjoy sunflower seeds and niger seeds.

Sighting the frostier-looking hoary redpoll, *Carduelis hornemanni,* among a flock of common redpolls was indeed a birding triumph for our friends Tom and Janet Rost. The bird is rarely seen south of Canada, even in winter.

The pine siskin, *Carduelis pinus,* a heavily-streaked small brown bird, is best identified by its two bright-yellow wing bars. It breeds mainly in Canada, the West, and Wisconsin to Maine. It winters in most of the lower 48 states. Pine siskins are reluctant to give up a position on the niger seed feeder, and the closer spring gets, the feistier they seem to become.

The American goldfinch, *Carduelis tristis*, bright canary yellow with jaunty black cap and wings, is a bird that is found throughout most of the country in winter. It readily accepts sunflower seeds, but is especially fond of niger seeds. In February, we see its olive-drab winter plumage start to show patches here and there of the brilliant yellow that will adorn the bird in summer months.

In February, the goldfinch's olive-drab winter plumage starts to show patches of its brilliant yellow breeding plumage.

Its breeding range extends from the central states northward into much of Canada. Late nesters, most goldfinches don't start building nests until July or August. The nest cup of woven plant fibers, thistle and cattail down is so tightly woven that it will actually hold water.

TOWHEE

A rustling in the dry leaves along our fencerow and a distinct *tow-hee!* are the trademarks of the rufous-sided towhee, *Pipilo erythrophthalmus.* A large, flashy bird, the male has a black head, throat, back and tail, bright white breast and rufous sides below the wings. The female is similarly marked, but with brown instead of black.

In the northern states, towhees are present only in the breeding season. In much of the rest of the country, towhees are year-round residents.

They are not commonly attracted to bird feeders, but towhees do enjoy using water areas for drinking and bathing.

SPARROWS

With the exception of the English, or house, sparrow (which isn't a sparrow at all, but a weaver finch), sparrows are lovely little birds that have traits we admire. They are clean and industrious, consume many insect pests and get along well with their fellow birds and man.

Dark-eyed junco, *Junco hyemalis,* is the name now used for the juncos that were formerly known as the white-winged, slate-colored, Oregon and gray-headed juncos. Sometimes called snowbirds, juncos are among the first of the avian winter residents to arrive in September or early October, and the last to migrate to their northern breeding grounds in spring.

They come each day for the seeds we provide at our feeders, especially the cracked corn on platform feeders or on the ground. Most juncos sport a gray vest, sharply cut off by an abrupt line where the white underparts begin. White outer tail feathers flash on and off as the birds fan their tails in flight. In the West, the juncos known as the Oregon race have a black hood, brown back and wings and white underparts.

The American tree sparrow, *Spizella arborea*, is another winter visitor whose breeding grounds are in the far northern areas of Canada and Alaska. It is a gentle, bright, energetic bird that is a regular at many winter feeding stations when it is not foraging the landscape for weed seeds. Years ago, one authority claimed that in the state of Iowa alone, tree sparrows consumed 875 tons of weed seeds each winter.

If you have a backyard of average size, you probably have chipping sparrows, *Spizella passerina*, each summer. They are commonly seen on lawns and seldom seem to be far from human

The American tree sparrow is a winter visitor whose breeding grounds are in the far northern areas of Canada and Alaska.

habitation. It is an unassuming little bird, about 5 inches long, that never draws attention to itself except when it sings its loud, musical trill. Its dainty nest is a cup of grasses and rootlets lined with hair and built on a low branch in a tree or hedge. Its three or four tiny eggs are bluish green with black spots.

We often hear white-throated sparrows, *Zonotrichia albicollis*, in our garden as they migrate through our area in spring and autumn, and occasionally see them at that time on one of our bird feeders. Their *Old Sam Peabody Peabody Peabody* song is a joy to hear, like a cheery greeting from an old friend. Whitethroats breed in Canada and from Wisconsin and Michigan to New England, as far south as Massachusetts and Connecticut. In the winter, they are found along the West Coast and in the southern and eastern part of the United States from Arizona to New England.

The song sparrow, *Melospiza melodia*, probably enjoys the widest distribution of any native North American bird. It ranges from Florida and Mexico in the winter to Alaska in the summer. It is a heavily-streaked brown sparrow that pumps its tail in flight. Its melodious, lively song belies its drab appearance. Its nest is usually on the ground, under a bush or tuft of grass. The cup of grasses, leaves and bark holds three to five greenish-white eggs that are heavily spotted with reddish brown. In the winter, song sparrows are steady feeding-station customers, eating amiably with the tree sparrows, juncos and whitethroats.

. . . K.P.H.

FOR MORE INFORMATION

Full chapters on the robin, cardinal, chickadee, blue jay, mourning dove, American goldfinch, mockingbird, house wren, white-breasted nuthatch and downy woodpecker are included in our book *America's Favorite Backyard Birds* (Simon and Schuster, 1983).

GLOSSARY

Aestivation: Torpor or dormancy during periods of extreme heat or drought.

Altruism: Behavior that benefits others at risk to oneself.

Amplexus: A wrapping around, or embrace (Latin).

Anthropomorphism: Attribution of human traits or behavior to animals.

Barking off: In 1828, Aubudon described "barking off squirrels" as a common way of killing squirrels in which a gunshot was fired to hit the bark of the tree immediately beneath the squirrel. Supposedly, the resulting concussion instantly killed the squirrel but did not mutilate it, so there was no wasted meat. Heavy barrels with small bores were used to fire the balls from these old firearms.

Binocular: Both eyes seeing the same thing; stereoscopic vision. Also see: Monocular.

Brood: The young in a single litter or a single hatching.

Browse: Twigs, leaves and bark on trees and shrubs eaten by some animals.

Browsing: The act of feeding on twigs, leaves or bark of trees and shrubs.

Brush: A luxurious, bushy tail, especially that of the fox.

Cache: Store in a hiding place.

Carnivorous: The term applied to those whose diet is mainly composed of meat or flesh.

Cementum: The hard, bony substance forming the outer layer of the root of a tooth.

Chorionic: Pertaining to the membrane which encloses the embryo of mammals, birds and reptiles.

Chrysalis: The third stage in the life of some insects, most notably butterflies and moths, in which it is enclosed in a hard outer shell.

Circadian rhythm: the body's own 24-hour timetable.

Coniferous: Pertaining to those trees which bear cones, such as pines, spruces, firs and cedars.

Coprophagy: Eating one's own excrement.

Copulation: Sexual union; coupling.

Covert: In birds, a feather which extends over or covers the base of the main feathers in the tail or wings.

Crepuscular: Active from dusk to dawn.

Deciduous: Pertaining to those trees which drop their leaves in autumn.

Digitigrade: An animal that walks on its toes, rather than on the entire sole of the foot. Cats, horses and dogs are examples of digitigrades.

Diurnal: Most active during daylight.

Embryonic: Pertaining to the embryo.

Estrus: The period of being in heat, when the female of a species is ready to breed.

Fecundity: The capacity to produce many offspring.

Forage: To wander in search of food.

Form: The molded, matted-down area in which an animal hides and sleeps. Cottontail rabbits and box turtles use forms.

Gestation: Pregnancy.

Grizzled: Having fur or hair which has a gray or silvery cast. Grizzly.

Heat: In animals, the period during which the female is ready to breed. Estrus.

Herbivorous: Pertaining to those that feed primarily on plants. Vegetarian.

Home range: The area in which an individual animal lives.

Homoiotherm: One that is warm-blooded, able to maintain a constant body temperature regardless of ambient temperature.

Larva: The wormlike form of an insect just hatched from the egg.

Malocclusion: A condition in which the upper and lower teeth do not meet properly. Overbite or underbite.

Marsupial: A mammal in which the young are born in an undeveloped state, then crawl into the external abdominal pouch, or marsupium, of the mother to complete their development.

Marsupium: The external abdominal pouch of a female marsupial in which she nurtures her young until they are fully developed.

Metamorphosis: A transformation during the growth process in which physiological changes occur in structure and habits.

Monocular: Each eye seeing independently. Also see: Binocular.

Monogamous: Having one mate.

Musk: An oily, glandular secretion, usually quite pungent.

Nocturnal: Most active at night.

Omnivorous: Pertains to those that feed on vegetable as well as animal matter.

Parotoid gland: In frogs and toads, one of a pair of large glands located on the head.

Pelage: An animal's coat of fur or hair.

Petagium: The loose, fur-covered flap of skin which extends from the wrist of the foreleg to the ankle of the hind leg on gliding animals such as the flying squirrel.

Pheromone: A substance secreted by an animal or insect as a communication, especially as a sexual attractant, to others of its kind.

Placenta: Afterbirth. The nutrient-rich, membranous organ that lines the uterine wall, nourishes the fetus, and is expelled following birth.

Plantigrade: One that walks on the entire sole of the foot. Man, raccoons, bears and skunks are examples of plantigrades.

Poikilotherm: One that is cold-blooded, whose body temperature rises or falls in response to the ambient temperature.

Polygamous: Having more than one mate at a time.

Predator: One that survives by killing and eating other species.

Prehensile: Capable of holding or wrapping around an object.

Pupa: The inactive stage in an insect's life between larva and adult.

Sac: A pouch, usually filled with fluid.

Spinneret: In some insects, a structure which excretes a silky filament to form webs or cocoons.

Torpid: Lethargic, inactive.

Vibrissae: Whiskers.

Volplane: To glide earthward.

REFERENCES

Allen, E. G. *The Habits and Life History of the Eastern Chipmunk (Tamias striatus lysteri)*. New York State Museum Bulletin 314, 1938.

Allen, J. A. "Catalogue of the Reptiles and Batrachians found in the Vicinity of Springfield, Massachusetts, with Notices of all the other species known to inhabit the state." *Proceedings of the Boston Society of Natural History*, 12:171–204, 248–50, 1868.

Allen, Thomas B., ed. *Wild Animals of North America*. Washington: National Geographic Society, 1979.

Allen, Thomas B.; Jensen, Karen; and Kopper, Philip. *Earth's Amazing Animals*. Washington: National Wildlife Federation, 1983.

Anon. "Butterfly Garden." *Ranger Rick's Nature Magazine*, May 1977, p. 33.

———. "Skunk Squirt," *Science Digest*, April 1975, p. 12.

Anthony, H. E., and McSpadden, J. Walker, eds. *Animals of America*. Garden City, New York: Doubleday, Doran & Company, Inc., 1937.

Arnett, Ross H., Jr., and Jacques, Richard L., Jr. *Simon and Schuster's Guide to Insects*. New York: Simon and Schuster, 1981.

Audubon, John J., and Bachman, Rev. John. *Quadrupeds of North America*. New York: George R. Lockwood, ca. 1849.

Bailey, Vernon. *Biological Survey of Texas*. 1905.

Barash, David P. "Marmot Alarm-Calling and the Question of Altruistic Behavior." *The American Midland Naturalist*, Vol. 94, No. 2, 1975, pp. 468–470.

Barkalow, Frederick, S., Jr., and Shorten, Monica. *The World of the Gray Squirrel*. Philadelphia: J. B. Lippincott Company, 1973.

Barkalow, F. S., Jr., and Soots, R. F., Jr. "Life Span and Reproductive Longevity of the Gray Squirrel, *Sciurus c. carolinensis* Gmelin." *Journal of Mammalogy*, Vol. 56, March 1975, pp. 572–574.

Barkalow, F. S.; Hamilton, R. B.; and Soots, R. F. "The Vital Statistics of an Unexploited Gray Squirrel Population." *Journal of Wildlife Management*, Vol. 34, No. 3, July 1970, pp. 489–500.

Barker, Will. *Familiar Animals of America*. New York: Harper & Brothers, 1951.

———. *Winter-Sleeping Wildlife*. New York: Harper & Brothers, 1958.

Barry, William J. "Environmental Effects on Food Hoarding in Deermice *(Peromyscus)*." *Journal of Mammalogy*, Vol. 57, No. 4, 1976, pp. 731–746.

Brower, Lincoln P., and Brower, Jan Van Zandt. "The Relative Abundance of Model and Mimic Butterflies in National Populations of the *Battus philenor* Mimicry Complex." *Ecology*, Vol. 43, No. 1, pp. 154–158.

Burroughs, John. *Squirrels and Other Fur-bearers*. Boston: Houghton Mifflin Company, 1900.

———. *The Writings of John Burroughs*. Vol. IX: *Riverby*. Boston: Houghton Mifflin Company/Riverside Press, 1894.

Burt, William H. *A Field Guide to the Mammals*. 3rd ed. Boston: Houghton Mifflin Company, 1976.

Cahalane, V. H. *Mammals of North America*. New York: Macmillan Company, 1947.

Camazine, Dr. Scott. *Chemical Ecology*, January 1984.

Colby, Constance Taber. "Warty Toads." *Country Journal*, July 1980, pp. 60–65.

Conant, Roger. *A Field Guide to Reptiles and Amphibians of Eastern and Central North America*. Boston: Houghton Mifflin Company, 1975.

Connolly, Michael S. "Time-Tables in Home Range Usage by Gray Squirrels." *Journal of Mammalogy*, Vol. 60, November 1979, pp. 814–817.

Cowan, Robert C. "What Keeps Some Frogs Warm in Freezing Temperatures? Antifreeze." *Christian Science Monitor*, February 17, 1982.

Davis, David E. "Mechanisms for Decline in a Woodchuck Population." *Journal of Wildlife Management*, Vol. 45, No. 3, July 1981, pp. 658–668.

———. "Role of Ambient Temperature in Emergence of Woodchucks *(Marmota monax)* from Hibernation." *American Midland Naturalist*, Vol. 97, No. 1, 1977, pp. 224–229.

DeGraaf, Richard M.; Witman, Gretchin M.; and Rudis, Deborah. *Forest Habitat for Mammals of the Northeast*. Amherst, Mass.: Forest Service, U. S. Department of Agriculture, 1981.

DeGraaf, Richard M., and Rudis, Deborah D. *Amphibians and Reptiles of New England*. Amherst, Mass: University of Massachusetts Press, 1983.

Devoe, Alan. *Our Animal Neighbors*. New York: McGraw-Hill Book Company, Inc., 1953.

———. *Speaking of Animals*. New York: Creative Age Press, 1947.

Doebel, John H., and McGinnes, Burd S. "Home Range and Activity of a Gray Squirrel Population." *Journal of Wildlife Management*, Vol. 38, October 1974, pp. 860–867.

Dolan, P. G., and Carter, D. C. "*Glaucomys volans*," *Mammalian Species*, No. 78, pp. 1–6, 1977.

Doutt, J. Kenneth; Heppenstall, Caroline A.; and Guilday, John E. *Mammals of Pennsylvania*. Harrisburg: Pennsylvania Game Commission in cooperation with Carnegie Museum, Carnegie Institute, Pittsburgh, 1966.

Drickamer, Lee C., and Capone, Michael R. "Weather Parameters, Trappability and Niche Separation in Two Sympatric Species of Peromyscus." *American Midland Naturalist*, Vol. 98, No. 2, 1977, pp. 376–381.

Ellis, Mel. "Some Findings to Nibble On." *Milwaukee Journal Insight*, June 28, 1981, p. 5.

Fafarman, Keith R., and Baker, Bruce W. "Hunger and Activity Levels of Cottontail Rabbits." *Journal of Mammalogy*, Vol. 60, No. 1, February 1979, pp. 212–213.

Fairbairn, Daphne J. "Why Breed Early? A Study of Reproductive Tactics in *Peromyscus*." *Canadian Journal of Zoology*, Vol. 55, 1977, pp. 862–871.

Fergus, Chuck. "Thornapples." *Pennsylvania Game News*, August 1983, pp. 50–53.

Fitzsimmons, Mark, and Weeks, Harmon P., Jr. "Observations on Snow Tunneling by *Sylvilagus floridanus*." *Journal of Mammalogy*, Vol. 62, No. 1, February 1981, pp. 211–212.

Forbes, S. E.; Lang, L. M.; Liscinsky, S. A.; and Roberts, H. A. *The White-Tailed Deer in Pennsylvania*. Harrisburg: Pennsylvania Game Commission, 1971.

Fritzell, Erik K. "Dissolution of Raccoon Sibling Bonds." *Journal of Mammalogy*, Vol. 58, No. 3, August 1977, pp. 427–428.

Garland, Lawrence, ed. "Deer Are Ruminants." *Habitat Highlights*, Vol. 4, No. 2, Spring 1984, p. 1.

———. "Seasonal Energy Cycle of Deer." *Habitat Highlights*, Vol. 4, No. 2, Spring 1984, pp. 1–3.

George, Jean. "The Delightful Delinquents." *National Wildlife*, Vol. 3, No. 3, April/May 1965, pp. 38–41.

Getty, Thomas. "Territorial Behavior of Eastern Chipmunks *(Tamias striatus)*: Encounter Avoidance and Spatial Time-Sharing." *Ecology*, Vol. 62, No. 4, 1981, pp. 915–921.

Gilbert, Bil. "Who Can Resist a Raccoon?" *Reader's Digest*, Vol. 114, May 1979, pp. 225–232.

Gilbert, F. E. *Raccoon*. Ottawa: Canadian Wildlife Service, 1975.

Giles, LeRoy W. "Food Habits of the Raccoon in Eastern Iowa." *Journal of Wildlife Management*, Vol. 4, No. 4, October 1940, pp. 375–382.

Gillette, Laurence N. "Movement Patterns of Radio-Tagged Opossums in Wisconsin," *American Midland Naturalist*, Vol. 104, No. 1, pp. 1–12.

Glaser, Harriet and Lustick, Sheldon. "Energetics and Nesting Behavior of the Northern White-Footed Mouse, *Peromyscus leucopus noveboracensis*." *Physiological Zoology*, Vol. 48, No. 2, April 1975, pp. 105–113.

Goodden, Robert. *The Wonderful World of Butterflies and Moths*. London: Hamlyn Publishing Group Ltd., 1977.

Goode, John. *Turtles, Tortoises, and Terrapins*. New York: Charles Scribner's Sons, 1971.

Gosling, Nancy Wells. "Flying Squirrels." *The Conservationist*, Vol. 36, May/June 1982, pp. 35–38.

Greenwood, Raymond J. "Nocturnal Activity and Foraging of Prairie Raccoons *(Procyon lotor)* in North Dakota." *American Midland Naturalist*, Vol. 107, No. 2, 1982, pp. 238–243.

Groves, John D. "Mass Predation on a Population of the American Toad, *Bufo americanus.*" *American Midland Naturalist,* Vol. 103, No. 1, 1980, pp. 202–203.

Gustafson, Anita. "Toads Are Terrific." *Ranger Rick's Nature Magazine,* May 1981, pp. 42–47.

Haitch, Richard. "Frogs' Tongues." *New York Times,* September 18, 1983.

Hamilton, William J. *The Mammals of the Eastern United States.* Ithaca, New York: Comstock. 1943.

Handley, Charles O., Jr., and Patton, Clyde. *Wild Mammals of Virginia.* Richmond, Virginia: Commonwealth of Virginia Commission of Game and Inland Fisheries, 1947.

Harrison, George. "Up to Our Hips in Chipmunks." *Sports Afield,* Vol. 180, No. 1, July 1978, p. 20

———. "They Walk on Snow." *Sports Afield,* December 1984.

Harrison, Hal H. *A Field Guide to Birds' Nests.* Boston: Houghton Mifflin Company, 1975.

———. *A Field Guide to Western Birds' Nests.* Boston: Houghton Mifflin Company, 1979.

———. *American Birds in Color.* New York: Wm. H. Wise & Co., Inc., 1948.

———. *Pennsylvania Reptiles & Amphibians.* Harrisburg: Pennsylvania Fish Commission, 1950.

Harrison, Kit. "Hemlock Gives Deer Best Winter Cover." *Sports Afield,* November 1981, p. 20.

———. "Metabolism Studies Help Game Managers." *Sports Afield,* November 1982, p. 22.

———. "About That Rabbit in Your Garden . . ." *Exclusively Yours,* May 12, 1980, pp. 76–81.

Harrison, Kit and George. *America's Favorite Backyard Birds.* New York: Simon and Schuster, 1983.

Hartman, Carl G. "Studies in the Development of the Opossum *(Didelphys virginiana).*" *Anatomical Record,* October 1920, pp. 251–261.

Heinold, Laura R. "Flying Squirrels—Treetop Gliders." *Science Digest,* September 1973, pp. 36–38.

Henisch, B. A. and H. K. *Chipmunk Portrait.* State College, Pennsylvania: Carnation Press, 1970.

Hirth, David H., and McCullough, Dale R. "Evolution of Alarm Signals in Ungulates with Special Reference to White-tailed Deer." *American Naturalist,* Vol. 111, No. 977, January/February 1977, pp. 31–42.

Hoffman, Cliff O., and Gottschang, Jack L. "Numbers, Distribution and Movements of a Raccoon Population in a Suburban Residential Community." *Journal of Mammalogy,* Vol. 58, No. 4, November 1977, pp. 623–635.

Holler, Nicholas R., and Marsden, Halsey. "Onset of Evening Activity of Swamp Rabbits and Cottontails in Relation to Sunset." *Journal of Wildlife Management,* Vol. 34, April 1970, p. 349.

Horner, Kent. "Home Range of the White-tailed Deer." *Deer & Deer Hunting,* August 1983, p. 55–58.

————. "The Role of the Doe." *Deer & Deer Hunting,* February 1984, pp. 36–41.

Houseknecht, Clyde R., and Tester, John R. "Denning Habits of Striped Skunks *(Mephitis mephitis)." American Midland Naturalist,* Vol. 100, No. 2, pp. 424–429.

Jackson, Hartley H. T. *Mammals of Wisconsin.* Madison, Wisconsin: University of Wisconsin Press, 1961.

Jones, J. K.; Carter, D. C.; Genoways, H. H.; Hoffman, R. S.; and Rice, D. W. *The Revised Checklist of North American Mammals North of Mexico.* Lubbock: Texas Tech University, 1982.

Kammermeyer, K. E., and Marchinton, R. L. "Notes on Dispersal of Male White-tailed Deer." *Journal of Mammalogy,* Vol. 57, No. 4, November 1976, pp. 776–778.

Keefe, Jim. *The World of the Opossum.* Philadelphia: J. B. Lippincott Company, 1967.

Kelsall, J. P. *Woodchuck.* Ottawa: Canadian Wildlife Service, 1973.

Kemmerer, Jack B. "City of the White Squirrels." *National Wildlife,* Vol. 6, No. 2, February/March 1968, pp. 30–31.

Kilpatrick, C. William; Kilpatrick, James W.; and Kilpatrick, Scott M. "A Wood-Mouse *(Peromyscus leucopus)* Nest Site in a Hornet's Nest." *American Midland Naturalist,* Vol. 105, No. 1, 1981, p. 208.

King, F. H. "Instinct and Memory Exhibited by the Flying Squirrel in Confinement." *The American Naturalist,* Vol. 17, No. 1, pp. 36–42, January, 1883.

Klots, Alexander. *A Field Guide to the Butterflies.* Boston: Houghton Mifflin Company, 1951.

Kriz, John J., "The Ups and Downs of Rabbits." *Pennsylvania Game News,* May 1971.

Lampe, David. ". . . He Has a Way with Squirrels." *National Wildlife,* Vol. 13, No. 6, October/November 1976, pp. 34–37.

Laycock, George. "Stupid, Ugly, Successful." *Audubon,* July 1983, p. 20.

Lewis, Allen R. "Selection of Nuts by Gray Squirrels and Optimal Foraging Theory." *American Midland Naturalist,* Vol. 107, April 1982, pp. 250–257.

Linzey, Donald W., and Linzey, Alicia V. "Growth and Development of the Southern Flying Squirrel *(Glaucomys volans volans)." Journal of Mammalogy,* Vol. 60, No. 3, August 20, 1979, pp. 615–620.

Lishak, Robert S. "Gray Squirrel Mating Calls: A Spectrographic and Ontogenic Analysis." *Journal of Mammalogy,* November 1982, pp. 661–663.

————. "Vocalizations of Nesting Gray Squirrels." *Journal of Mammalogy,* Vol. 63, August 1982, pp. 446–452.

Lockley, R. M. *The Private Life of the Rabbit.* New York: Macmillan Publishing Co., Inc., 1964.

Long, T. A.; Cowan, R. L.; Wolfe, C. W.; Rader, Terry; and Swift, R. W. *Effect of Seasonal Feed Restriction on Antler Development of White-tailed Deer.* University Park, Pennsylvania: Pennsylvania State University College of Agriculture, 1959.

MacArthur, Kenneth. "A False Front." *Museum Record,* January 1948, p. 22.

MacClintock, Dorcas. *The Squirrels of North America.* New York: Van Nostrand Reinhold Company, 1970.

Madson, John. *The Cottontail Rabbit.* 2nd ed. East Alton, Ill,: Olin Mathieson Chemical Corporation, 1959.

——. *Gray and Fox Squirrels.* East Alton, Ill.: Olin Mathieson Chemical Corporation, 1964.

——. "One Giant Leap for Deerkind." *Audubon,* Vol. 81, September 1979, pp. 106–107.

——. *The White-tailed Deer.* East Alton, Ill.: Olin Mathieson Chemical Corporation/Winchester-Western Press, 1961.

Magruder, N. D.; French, C. E.; McEwen, L. C.; and Swift, R. W. *Nutritional Requirements of White-tailed Deer for Growth and Antler Development II.* University Park, Penn.: Pennsylvania State University College of Agriculture, 1957.

May, Charles Paul. *Box Turtle Lives in Armor.* New York: Holiday House, 1960.

McDonald, Joe. "Peeper Fever." *Animal Kingdom,* Vol. 84, No. 1, February/March 1981, pp. 4–9.

McDonough, James J. *The Cottontail in Massachusetts.* Boston: Massachusetts Division of Fisheries and Game, undated.

McManus, J. J. "Behavior of Captive Opossums." *American Midland Naturalist,* Vol. 84, No. 1, July 1970, pp. 144–169.

Merriam, C. Hart. *Mammals of the Adirondack Region.* New York: Published by the author. 1884.

Miller, R. S. *Striped Skunks.* Ottawa: Canadian Wildlife Service, 1973.

Minton, S. A, Jr. "Introduction to the Study of the Reptiles of Indiana." *American Midland Naturalist,* Vol. 32, 1944, pp. 438–477.

——. *Amphibians and Reptiles of Indiana.* Indianapolis: Indiana Academy of Science, 1972.

Mitchell, Joseph C. "Frogs and Toads of Virginia." *Virginia Wildlife,* April 1975, pp. 13+.

Mitchell, Robert T., and Zim, Herbert S. *Butterflies and Moths.* New York: Golden Press/Western Publishing Company, Inc., 1962.

Neal, Jay T. "Elves That Fly by Night." *National Wildlife,* December/January 1967, p. 36.

Newsom-Brighton, Maryanne. "Butterflies Are Free." *National Wildlife,* Vol. 20, No. 3, April/May 1982, pp. 27–28+.

Nicholls, Richard E. *The Running Press Book of Turtles.* Philadelphia: Running Press, 1977.

Nixon, Charles M. "Insects as Food for Juvenile Gray Squirrels." *American Midland Naturalist,* Vol. 84, July 1970, p. 283.

Oldham, R. S. "Spring Movements in the American Toad, *Bufo americanus.*" *Canadian Journal of Zoology,* Vol. 44, 1966, pp. 63–100.

Packard, Fred M. "Tufted Titmice pull hairs from living mammals." *Journal of Mammalogy.* 30(4):432. Nov. 21, 1949.

Palmer, E. Laurence. *Fieldbook of Mammals.* New York: E. P. Dutton & Company, Inc., 1957.

Palmer, Ralph S. *The Mammal Guide.* Garden City, N. Y.: Doubleday & Company, Inc., 1954.

Passmore, R. C. *White-tailed Deer.* Ottawa: Canadian Wildlife Service, 1973.

Patent, Dorothy Hinshaw. *Butterflies and Moths: How They Function.* New York: Holiday House, 1979.

Peterson, Barbara and Russell F. *Whitefoot Mouse.* New York: Holiday House, 1959.

Peterson, Roger Tory. *A Field Guide to the Birds.* 4th ed. Boston: Houghton Mifflin Company, 1980.

———. *A Field Guide to Western Birds.* 2nd ed. Boston: Houghton Mifflin Company, 1969.

Porter, George. *The World of the Frog and the Toad.* Philadelphia: J. B. Lippincott Company, 1967.

Rachesky, Stanley. "*Nothing* Repels Rabbits Bent on Cabbage." *Flower and Garden,* September 1979, p. 47.

Rawlins, John Edward. "Thermoregulation by the Black Swallowtail Butterfly, *Papilio polyxenes* (Lepidoptera: Papilonidae)." *Ecology,* Vol. 61, No. 2, pp. 345–357.

Robbins, Chandler S.; Bruun, Bertel; and Zim, Herbert S. *Birds of North America.* New York: Golden Press/Western Publishing Company, 1983.

Robinson, Wirt. "Woodchucks and Chipmunks." *Journal of Mammalogy,* 4(4):256–57, Nov. 1, 1923.

Rose, George B. "Mortality Rates of Tagged Adult Cottontail Rabbits." *Journal of Wildlife Management,* Vol. 41, No. 3, July 1977, pp. 511–514.

Rue, Leonard Lee, III. *The Deer of North America.* New York: Crown Publishers, Inc., 1978.

———. *Sportsman's Guide to Game Animals.* New York: Outdoor Life Books, 1968.

———. *Cottontail.* New York: Thomas Y. Crowell Company, 1965.

———. *The World of the Raccoon.* Philadelphia: J. B. Lippincott Company, 1964.

———. "Whitetail Deer." *American Hunter,* September 1984, pp. 42–43+.

———. *The World of the White-tailed Deer.* Philadelphia: J. B. Lippincott Company, 1962.

Sadler, Ken. "Of Rabbits and Habitat . . . A Long-Term Look." *Pennsylvania Game News,* October 1980, pp. 7–11.

Sadler, Kenneth C. "Common Rabbit Diseases & Parasites." *Missouri Conservationist,* November 1983, pp. 21–25.

Sanderson, Glen C. "Raccoon Values—Positive and Negative." *Illinois Wildlife,* Vol. 16, No. 1, December 1960.

Sandved, Kjell B., and Emsley, Michael G. *Butterfly Magic.* New York: Viking Press, Inc., 1975.

Schmid, William D. *Science,* February 5, 1982.

Schoonmaker, W. J. *The World of the Woodchuck.* Philadelphia: J. B. Lippincott Company, 1966.

Schwartz, Elizabeth; Schwartz, Charles; and Kiester, A. Ross. *The Three-Toed Box Turtle in Central Missouri, Part II: A Nineteen-Year Study of Home Range, Movements and Population.* Jefferson City, Mo.: Missouri Department of Conservation, 1984.

Scott, Jack Denton. "Cheerful Chatterbox the Chipmunk." *National Wild-*

life, Vol. 11, No. 3, April/May 1973, pp. 22–24.

Scott, James A. "Mate-Locating Behavior of Butterflies." *American Midland Naturalist*, Vol. 91, No. 1, 1974, pp. 103–117.

Scott, Shirley L., ed. *Field Guide to the Birds of North America*. Washington: National Geographic Society, 1983.

Seton, Ernest Thompson. *Life-Histories of Northern Animals*. 2 Vols. New York: Charles Scribner's Sons, 1909.

––––––. *Lives of Game Animals*. 4 Vols. Garden City, N. Y.: Doubleday, Page and Company, 1925.

Severey, Merle, ed. *Wild Animals of North America*. Washington: National Geographic Society, 1960.

Shaffer, Larry. "Use of Scatterhoards by Eastern Chipmunks to Replace Stolen Food." *Journal of Mammalogy*, Vol. 61, No. 4, November 1980, pp. 733–734.

Sims, Steven R."Aspects of Mating Frequency and Reproductive Maturity in *Papilio zelicaon*." *American Midland Naturalist*, Vol. 102, No. 1, 1979, pp. 37–49.

Smith, Luther. "Observations on the nest-building behavior of the opossum." *Journal of Mammalogy*, 22(2):201–2, May 13, 1941.

Smith, Ned. "Waddling Wonders of the Wood Lot." *Pennsylvania Game News*, January 1962, pp. 4–6.

Stack, J. W. *Journal of Mammalogy*, May 1925, pp. 128–129.

Sterba, James P. "Frog's Tongue: The Secret of Its Speed." *New York Times*, June 15, 1982.

Stickel, L. F. "Population and Home Range Relationships of the Box Turtle, *Terrapene c. carolina* (Linnaeus)." *Ecology Monograph*, Vol. 20, 1950, pp. 351–378.

Stoddard, H. L. "The Flying Squirrel as a Bird Killer." *Journal of Mammalogy*, 1(2):95–96, March 2, 1920.

Sunquist, M. E. "Winter Activity of Striped Skunks *(Mephitis mephitis)* in East-Central Minnesota." *American Midland Naturalist*, Vol. 92, No. 2, 1974, pp. 434–445.

Swihart, Robert K., and Yahner, Richard H. "Browse Preferences of Jackrabbits and Cottontails for Species Used in Shelterbelt Plantings." *Journal of Forestry*, February 1983, pp. 92–93.

Taylor, Walter P., ed. *The Deer of North America*. Harrisburg and Washington: Stackpole Company and the Wildlife Management Institute, 1956.

Thomas, Kim Rutherford. "Burrow Systems of the Eastern Chipmunk *(Tamias striatus popilan Lowery)* in Louisiana." *Journal of Mammalogy*, Vol. 55, No. 2, May 1974, pp. 454–459.

Thompson, D. C. "Reproductive Behavior of the Gray Squirrel." *Canadian Journal of Zoology*, Vol. 55, July 1977, pp. 1176–1184.

––––––. "The Social System of the Gray Squirrel." *Behaviour*, Vol. 64, 1978, pp. 305–328.

Thompson, D. C., and Thompson, P. S. "Food Habits and Caching Behavior of Urban Squirrels." *Canadian Journal of Zoology*, Vol. 58, May 1980, pp. 701–710.

Trent, T. T., and Rongstad, O. H. "Home Range and Survival of Cottontail

Rabbits in Southwestern Wisconsin." *Journal of Wildlife Management,* Vol. 38, No. 3, July 1974, pp. 459–472.

Tyler, Hamilton A. *The Swallowtail Butterflies of North America.* Healdsburg, Calif.: Naturegraph Publishers, Inc., 1975.

Tyndale-Biscoe, C. H., and Mackenzie, R. B. "Reproduction in *Didelphis marsupialis* and *D. albiventris* in Colombia." *Journal of Mammalogy,* Vol. 57, No. 2, pp. 249–265.

Vance, Joel M. "The Return of Hairy Houdini." *Missouri Conservationist,* February 1984, pp. 12–15.

Vogt, Richard Carl. *Natural History of Amphibians and Reptiles in Wisconsin.* Milwaukee: Milwaukee Public Museum, 1981.

Waldman, Bruce, and Adler, Kraig. "Toad Tadpoles Associate Preferentially with Siblings." *Nature,* Vol. 282, December 6, 1979, pp. 611–613.

Walker, Ernest P. " 'Flying' Squirrels, Nature's Gliders." *National Geographic,* May 1947, pp. 663–674.

Wegner, Robert. *Deer & Deer Hunting.* Harrisburg: Stackpole Books, 1984.

Wernert, Susan J., ed. *North American Wildlife.* Pleasantville, N. Y.: Reader's Digest Association, 1982.

Whipple, A. B. C. "The Raccoon Life in Darkest Suburbia," *Smithsonian,* August 1979, pp. 83–87.

Whitfield, Dr. Philip, ed. *Macmillan Illustrated Animal Encyclopedia.* New York: Macmillan Publishing Company, 1984.

Williams, Ted. "Fairy Diddling." *Audubon,* Vol. 85, No. 6, November 1983, pp. 14–20.

Williamson, Lonnie. "Whitetails Make the Map." *Outdoor Life,* February 1984, pp. 24–26.

Wingard, R. B., and Studholme, C. R. *Eastern Chipmunks.* Special Circular 97. University Park, Penn.: Pennsylvania State University College of Agriculture, undated.

Wishner, Lawrence. *Eastern Chipmunks, Secrets of Their Solitary Lives.* Washington: Smithsonian Institution Press, 1982.

Yahner, Richard H. "Burrow System and Home Range Use by Eastern Chipmunks, *Tamias striatus:* Ecological and Behavioral Considerations." *Journal of Mammalogy,* Vol. 59, No. 2, May 1978, pp. 324–329.

Yahner, Richard H., and Svendsen, Gerald E. "Effects of Climate on the Circannual Rhythm of the Eastern Chipmunk, *Tamias striatus.*" *Journal of Mammalogy,* Vol. 59, No. 1, February 1978, pp. 109–117.

Zim, Herbert S., and Smith, Hobart M. *Reptiles and Amphibians.* New York: Golden Press/Western Publishing Company, Inc., 1953.

INDEX

CREDITS

(Continued from copyright page)

Karl Maslowski: pages 21, 45 (top), 51, 54, 57, 85, 92, 102, 106 *(right)*, 136, 158 *(top)*, 170, 196, 202, and 241.
Karl and Steve Maslowski: page 173 (top).
Steve Maslowski: pages 59, 62 *(top)*, 73, 77, 79, 82, 109, 113, 114, 148 *(right)*, 169 *(right)*, 175, and 222.
Leonard Lee Rue III: pages 20, 36, 40, 45 *(bottom)*, 52 *(top)*, 55 *(top)*, 56, 60, 78, 118, 126, 138, 148 *(left)*, 149, 153, 154, 163 *(left)*, 173 *(bottom)*, 177, 187, 207, 211, 236, 237, 259, 260, and 272.
Charles Schwartz: pages 80, 89, 104 (right), 132, and 140.
Charles and Elizabeth Schwartz: pages 55 *(bottom)*, 178, 224, 243, 246 (top), 248, 252 *(bottom)*, 265, and 267.
Irene Vandermolen (Leonard Lee Rue Enterprises): pages 74, 81, 83, 90, 115 *(left)*, 155, 157, 206, 209, 214, and 215.

Line drawings appear on the following pages:

Ned Smith: pages 3, 4, 5, 9, 19 (Pa. Game Comm.), 23 (Pa. Game Comm.), 39, 41 (Pa. Game Comm.), 65, 67 (Pa. Game Comm.), 83, 86, 101 (Pa. Game Comm.), 123, 125, 129 (F/M '69 *National Wildlife*), 143, 145 (Pa. Game Comm.), 161 (A/S '66 *National Wildlife*), 164, 165 (Pa. Game Comm.), 182, 183 (J/J '67 *National Wildlife*), 190 (J/J '67 *National Wildlife*), 200, 201 (Pa. Game Comm.), 205 (Pa. Game Comm.), 216 (D/J '73 *National Wildlife*), 217 (Pa. Game Comm.), 218, 219 (Pa. Game Comm.), 231 (D/J '67 *National Wildlife*), 233, 250, 251 (Pa. Game Comm.), 262 (Pa. Game Comm.), 275, 277 (J/J '70 *National Wildlife*), and 302.
Michael James Riddet: pages 87, 94, 98, 99, 235, 238, and 244.